The Urban Forest

The Urban Forest

COMPREHENSIVE MANAGEMENT

Gene W. Grey

JOHN WILEY & SONS, INC.

New York • Chichester • Brisbane • Toronto • Singapore

Copyright © 1996 by John Wiley & Sons, Inc.

Library of Congress Cataloging in Publication Data:
Grey, Gene W., 1931–
     The urban forest : comprehensive management / Gene W. Grey.
       p. cm.
     Includes bibliographical references (p.   ).
     ISBN 0-471-12275-0 (cloth : alk. paper)
     1. Urban forestry. I. Grey, Gene, 1931–   Urban forestry.
     II. Title.
     SB436.G72   1996
     635.9'77'091732—dc20   95-10875

Printed in the United States of America
10 9 8 7 6 5 4 3 2 1

To Jim Nighswonger, who taught me that urban forestry
is not only a science but also an art

Preface

O ver the years, much time has been spent (and perhaps wasted) in trying to define urban forestry. The task of definition is made difficult for a number of reasons, beginning with the necessity of accepting urban areas as forests and continuing with the need to understand the relationship of trees to other elements of the urban environment. The latter is critical to getting at urban forestry. I find it helpful to consider the physical urban environment as being made up of three basic (and dynamic) elements: trees and related organisms, structures (the trappings of society), and people. The interactions and relationships of these basic elements and all their appendages may properly be called the urban forest ecosystem.

I find it further helpful to consider that trees are, or should be, there to enhance the urban environment, and in so doing, they must be compatible and functional. Thus I arrive at a definition of urban forestry as that which must be done to make trees compatible and functional in the urban environment. I am comfortable with this definition, because the words "that which must be done" include not only doing something physically to or with trees, but they also involve planning, developing, funding, and whatever else leads to physically doing something.

A definition of urban forest management, which is what this book is all about, is also found in this definition of urban forestry. Management is simply the act of causing to happen those things that must be done to make trees compatible and functional in the urban environment.

Before continuing, it might be useful to consider whether trees are central to the urban environment. I think not. Trees are central to the urban forest but not to the total urban environment. It is currently popular to refer to trees as a component of the urban infrastructure, along with streets, utilities, sewers, and perhaps other structures. Such a reference may be of political value, but it must be tempered with the reality of the relative importance of trees in the urban environment. The word *compatible* in the definition of urban forestry suggests compromise.

The emphasis of this book is on comprehensive management, as exercised, ideally, by a central authority (city forestry department or other agency) with adequate knowledge, planning, and funding. The challenge of comprehensive management is to recognize the complexity of the total urban forest and to put in place and operate a system that meets the needs of trees and related organisms, structures, and people.

The book expands on five basic concepts: the entire urban forestry environment must be considered; six elements are necessary for comprehensive management; the needs of the total urban forest must be recognized; appropriate management must be applied, both directly and indirectly; and all management needs are interrelated, and all operations must be in context with long-range objectives.

For consistency, the terms *urban forest* and *urban forestry* are used throughout, even though I am aware that *community forest* and *community forestry* are sometimes preferred. Also, the terms *tree board, city forestry department,* and *city forester* are used, because the purpose here is to indicate the function rather than the various titles of these offices. Also, *urban forest manager* indicates a person, generally a representative of a city forestry department, with responsibility for specific functions. Additionally, *city* as used here refers to urban settings, whether small towns, villages, communities, large cities, or other units of local government with jurisdiction for urban forestry. The latter is quite important in that comprehensive management applies equally well to small rural villages and to large cities.

As indicated by the title, the text is intended as an introduction, or guide, with an emphasis on management rather than technical operations. I made a sincere effort to list major technical references for readers who may wish to dig deeper into specific topics.

Finally, and most importantly, the book is for students of urban forestry, for those who actually manage the forest, and for others whose interest in trees translates into making urban areas better places.

Acknowledgments

I extend my special thanks to Sandra Grey for her assistance, suggestions, and patience during preparation of the manuscript and for helping me with all the details of publication. I also thank those on whose work I have drawn and apologize to others whose relevant work I may have missed. I particularly appreciate The National Arbor Day Foundation for providing me a teaching forum for presenting the rudiments of comprehensive management. And, finally, I thank all the students, urban forest managers, and others who may find something of value in this work.

Gene W. Grey

Contents

PREFACE vii

Chapter

1 COMPREHENSIVE MANAGEMENT: THE CONCEPT AND
  REQUIREMENTS 1

2 A RESPONSIBLE ORGANIZATION 5

3 THE URBAN FORESTRY ENVIRONMENT 9

4 DETERMINING WHAT THE URBAN FOREST NEEDS 35

5 PLANNING AND BUDGETING FOR URBAN FORESTRY 45

6 PROGRAM IMPLEMENTATION 57

7 LEVERAGING YOUR EFFORTS 105

8 SUMMARY 129

Appendix

1 WESTMINISTER TREE COMMISSION ANNUAL
  REPORT FOR 1994 131

2 SUMMARY OF LONG-RANGE URBAN FORESTRY PLAN 139

3 A BASIC URBAN FORESTRY REFERENCE LIBRARY 143

4   FINANCIAL AND TECHNICAL HELP FOR URBAN FORESTRY
       PROGRAMS   145

5   HOW TO WRITE GRANTS—A STATE GRANTOR'S PERSPECTIVE   147

       SELECTED BIBLIOGRAPHY FOR URBAN FOREST INVENTORY   151

       INDEX   153

The Urban Forest

CHAPTER 1

Comprehensive Management:
The Concept and Requirements

THE CONCEPT

Fundamental to the concept of comprehensive management of the urban forest is visualizing the total urban area and understanding the complexities of location, ownership, and condition of the urban forest. Perhaps the best way to begin to understand the urban forest is from afar, starting with an overall view as from an aircraft or an aerial photograph. Here we will see nearly countless situations of tree-lined streets, highways nearly barren of trees, open areas, areas with occasional trees, and dense natural forests—all represented by a myriad of owners and each falling within a particular land use category. It is important that we view the scene in its entirety and that we recognize all areas, tree-covered and open, and all ownerships and situations. The scene is also dynamic. Trees grow and changes occur. At nearly all times, somewhere within the urban forest, trees are being planted, removed, pruned, or otherwise treated or mistreated—a bit here and a bit there, according to the needs and desires of a complex society. For example, on any given day at least one (and perhaps more) of the following events might occur: traffic-obstructing branches are removed by city crews, hazardous limbs are removed by a contracting arborist, shade trees are planted in a neighborhood park by local volunteers, trenches are dug near trees by a utility company contractor, overhanging branches are cut from a neighbor's tree by an irate homeowner, and trees are pruned, sprayed, and fertilized in dozens of backyards by homeowners and tree service companies.

We must recognize the totality of the urban forest and study each of its parts. We also must understand each part in terms of primary land use, ownership, physical (including biological) features, and anticipated changes. The various parts of the urban forest are examined in Chapter 3.

We also must understand the dynamic nature of the urban forest. It is not enough to simply know that trees will grow and that the urban landscape will change. It is essential that we appreciate all the factors of change—biological, social, political, and economic—within the urban environment and that we correctly anticipate change.

The concept of comprehensive management suggests orchestration— someone looking at the total picture and causing things to be done to influence the various parts of the urban forest. Because of the varied nature of the forest, particularly ownership, all parts cannot be treated equally. Hence, comprehensive management has two basic aspects: direct management (doing something directly to parts of the urban forest) and indirect management (influencing others to do something to other parts).

Who orchestrates management of the urban forest? Comprehensive management must be centrally authorized, vested in a single organization with responsibility for both direct and indirect management of the total forest.* Such an organization is usually a city forestry department, authorized by city codes, administratively responsible to city government, and often advised or guided by a citizens' board or commission.

On certain lands—generally those owned by city government or where easements have been given—the city forestry department exercises direct management, doing everything necessary to ensure the compatibility of trees with the urban environment. On large areas of the urban forest—mostly those in private ownership—the department has little direct authority, however, and must manage them indirectly, primarily through cooperation and educational efforts. In short, the city forestry department can plant, prune, remove, or do whatever else is desirable to certain parts of the urban forest, but it can only influence such practices on other parts. This is the single most important concept of comprehensive management.

---

*You must understand that central authority is for program orchestration only. In no way do I suggest that authority should be over other entities (such as private arboricultural firms) involved in urban forestry.

THE REQUIREMENTS

As stated previously, a responsible central organization is an essential element for comprehensive management of the urban forest. There are five other elements, each of which is necessary, for if any one is missing, management can be piecemeal at best. The six requirements for comprehensive management are:

A central organization with responsibility and authority

Knowledge of the total urban forestry environment—biological, institutional/social, and legal

Knowledge of what the urban forest needs

Plans for meeting the needs

Adequate budgets

Effective implementation

To emphasize their importance, I refer to these requirements as the six "gottas" of comprehensive urban forest management.

1. You gotta be responsible.
2. You gotta know the urban forestry environment.
3. You gotta know what the urban forest needs.
4. You gotta have a plan.
5. You gotta have money.
6. You gotta do it right.

To understand and appreciate comprehensive management, you need to put yourself mentally in the position of city forester, the someone responsible, and to think in terms of the total urban forestry environment. The requirements for comprehensive management and what goes into putting each in place are considered in the following chapters.

SUMMARY

Understanding the complexities of location, ownership, and condition of the urban forest is fundamental to comprehensive management. The dynamic nature of the forest and the factors of change also must be recognized. Comprehensive management requires orchestration by a central organization, which would apply management directly to certain parts and indirectly to other parts, according to ownership and other factors. There are five additional requirements: (1) knowledge of the total environment, (2) knowledge of urban forest needs, (3) plans, (4) budgets, and (5) effective implementation.

CHAPTER 2

A Responsible Organization

You gotta be responsible.

Someone must be responsible. Someone must orchestrate management of the total urban forest. Thus, there must be an organization with responsibility, authority, and effective leadership. The function must be that of a city forestry department with responsibility for betterment of the total urban forest within the city's jurisdiction. As indicated in Chapter 1, such responsibility is for direct management of certain segments but is limited to indirect management in other situations. The function of a city forestry department, particularly in larger cities, is often vested in an organization by that title but may also be carried out by other governmental units such as Public Works, Natural Resources, or Parks and Streetside Resources. In small cities lacking monetary resources for a separate city forestry department, care of the urban forest may be incorporated with other public works duties. For example, in the city of Westminster, Maryland, with a population of 14,600, leadership for urban forestry is combined with city planning. The city planner is also city forester, with training in both areas. The city forestry program is supported and directed by a citizens' tree commission. In Cincinnati, Ohio, by contrast, urban forestry is a division within Park Operations.

No matter how the city forestry program is organized, someone must have authority and responsibility. Authority is generally established by ordinance or charter, which establishes the departmental function and provides the legal basis for operations. In many cities, by ordinance, the authority of the city forester is only for trees on city property, with no stated responsibility for indirect management on other ownerships. In some cases, there may be implied responsibility. In nearly all cities, though, there are indirect management efforts by city forestry departments. Ideally, however, such responsibility should be clearly provided by ordinance. Chapter 3 discusses ordinances in more detail. Ultimately, ordinances can provide only the legal basis for individual performance. It is the city forester and staff, often supported or administered by a citizens' board, who must orchestrate management.

TREE BOARDS

Tree board is a generic title for citizens' groups with official support or administrative responsibilities for city forestry programs. Such groups may be known variously as commissions, committees, boards, or possibly other names. Established in most cases by ordinances, tree boards have varying official roles, which may be advisory, policy-making, administrative, or operational. Unofficially, tree boards often serve as urban forestry program advocates, generating public support and influencing budgets. Because responsibilities vary, clear distinctions are difficult; however, tree boards are generally classified as advisory, policy-making, or administrative.

Advisory boards are charged with giving counsel to city administrators concerning urban forestry matters. With no authority to expend funds or conduct operations, these groups study, investigate, and make recommendations. In so doing, they serve as representatives of the urban forest and its various owners and users. Thus, they often assume advocacy roles in building citizen support for urban forestry programs. In many cases, the city forester, or other person responsible for urban forest management, serves as an ad hoc member of the board.

Policy-making boards differ from advisory boards in that they generally are responsible for program planning, budgeting, and often implementation. They are not independent and must operate within the administrative framework of city government concerning operations and expenditure of funds. Such boards are most common in small cities lacking resources for forestry departments, and often assume the function of such departments. An excellent example is the McAlester, Oklahoma, tree board. McAlester, with a population of 16,000, does not have a city forestry department. Urban forestry is one of the many responsibilities of the city's land maintenance supervisor. "The all-volunteer tree board writes grants, organizes public relations events and Arbor Day ceremonies, coordinates environmental education programs, writes a weekly column in the newspaper, and organizes tree care seminars" (Westphal and Childs, 1994). Policy-making boards are not limited to small cities, however; some large cities have tree boards with planning and policy-making responsibilities.

Administrative boards are independent commissions with full urban forestry program responsibility and independent financing, often based on tax levies. Decision making concerning planning, budgeting, and management oversight is vested in the board of commissioners. Some commissions are not exclusive to urban forestry and may also be responsible for parks and recreation or other operations. Not unlike a corporate structure, there is generally a director or chief operating officer directly responsible to the board of commissioners. In cases of singular urban forestry commissions, this person is the city forester. Specific duties of administrative boards include employing the chief operating officer, strategic planning, policy development, budget approval, and overall program evaluation.

As stated previously, tree boards often assume an advocacy role. Rarely authorized by ordinance or charter, but sometimes implied, advocacy efforts can be extremely beneficial to urban forestry programs, resulting in citizen support and sustained funding. Advocacy takes many forms. It may be as subtle as speaking or writing about the benefits of trees or as direct as planned campaigns to influence budgets. From my observations, advocacy is a natural role for most tree board members. Members are interested in trees and are public-spirited individuals who are generally respected in their communities and who see their efforts on the tree board as a way of helping to improve their communities.

Tree board members speak for trees. They also represent the interests of citizens concerning trees. An additional role, which is synergistic with the city forestry department, is serving as an extension of the department, formally or informally, in providing information and education to individuals and groups concerning trees. Such activities help to gain public support for the total urban forestry program.

The value of active tree boards in comprehensive management is difficult to overstate. Please refer to Appendix 1. Reproduced in total is the 1994 annual report of the Westminister, Maryland, Tree Commission. This report reveals a wide scope of activities and illustrates clearly the value of tree boards to city forestry programs.

SUMMARY

Orchestration of management of the urban forest must be done by a central organization that has authority, takes responsibility, and provides effective leadership. The function is that of a city forestry department, but it may be vested in and carried out by other governmental units. Authority is established by ordinance or charter. Many city forestry departments are supported by tree boards whose various roles may be advisory, policy-making, or administrative. In small cities, tree boards may assume the operational role of the city forestry department. Citizen tree boards often assume an advocacy role, speaking for trees, resulting in sustained funding.

REFERENCE

Westphal, L., and G. Childs. 1994. Overcoming obstacles: Creating volunteer partnerships. *Journal of Forestry* 92(10):31.

CHAPTER 3

The Urban Forestry Environment

===

You gotta know the urban forestry environment.

Fundamental to comprehensive management is an understanding of the total environment in which urban forestry must operate. The total urban forestry environment includes all the physical (including biological), institutional, social, legal, and political factors that either facilitate or inhibit management.

THE PHYSICAL ENVIRONMENT

The physical environment refers to trees and related organisms and their various situations, including site, species, size, condition, location, ownership, and responsibility. Of these considerations, who is responsible is critically important, for these individuals directly influence, for better or worse, the total urban forest. It is also these individuals to whom educational efforts must be directed. Urban forest managers must understand the intended and actual use of the various areas of the urban forest and the benefits and liabilities of trees and other vegetation. They also must understand that the uses of the urban forest are dynamic; technology and social changes cause different public demands over time.

Those who are responsible—those who physically cause trees to be planted, pruned, removed, or otherwise treated—can be divided into three groups: city government, utility companies or contractors operating on easements, and other property owners. Thus, for comprehensive management, these three groups constitute three distinct audiences for the city forester.* More specific audiences can be identified within each group.

Each group is responsible for many different situations. As we consider each group, think in terms of primary land use, priorities, opportunities, and impediments to urban forest management.

Directly Manageable Urban Forest Lands

City government generally is responsible for urban forest management on the following lands: city parks, public squares, grounds of city buildings, monuments and cemeteries, boulevard medians, streetsides, parking lots, riparian zones, or other areas where special authority may be given. City forestry departments can generally apply direct management in these situations, although occasionally they may need to use indirect measures on lands managed by other units of city government.

City Parks.   City parks vary greatly, from active and passive areas with attendant structures to natural areas with minimal development. (See Figure 3.1.) Trees are nearly always present in parks as either natural forests, residual from natural forests, or planted specimens. There are also often large open areas, dedicated to active sports or other activities. Adjacent to such areas are often trees, strategically placed to provide shade, wind protection, screening, or other benefits. (See Figure 3.2.) Because of high public use, city parks require intensive management. Special measures must be taken to ensure survival and growth of newly planted trees, and all trees must be maintained for health and safety.

Public Squares.   Public squares might properly be classified as city parks in that they are generally city-owned green spaces of high public use. They differ somewhat, however, because of their relative small size and type of use. Generally located in the central city as town squares or other commons in smaller cities or as dedicated spaces in larger cities, such areas are often shaded gardens where people linger for conversation, lunch, perhaps chess or checkers, or other passive uses. (See Figure 3.3.) These areas might also be used for political demonstrations or, unfortunately, in modern society as centers for drug transactions or as refuges for the homeless. Thus, public squares

---

*As previously stated, the terms *city forester* or *city forestry department* convey the function rather than the specific title of the office and indicate responsibility for comprehensive management of the urban forest.

FIGURE 3.1. Parks may be either active or passive.

FIGURE 3.2. Strategically located, trees provide screening and shade in active park areas.

FIGURE 3.3. Trees are extremely important in outdoor malls and plazas, and require intensive care.

present special vegetation management problems. In many such areas, for example, shrubbery has been removed and trees have been pruned higher than normal to discourage criminal activity and provide easier surveillance by police.

Grounds of City Buildings. Grounds surrounding city buildings are important parts of the total urban forest. Often well landscaped with trees and other vegetation, such areas make positive contributions to the urban environment. (See Figure 3.4.) Included are city halls; buildings housing police, fire, and other departments; hospitals; schools; and buildings for other services. These grounds may also be used as passive parks, because they are often indistinguishable from parks or city squares.

Monuments and Cemeteries. Again, citizens frequently have difficulty clearly separating parks and city squares from monuments to individuals or events of historical significance because they share the same space. Many are on dedicated lands, however, and are thus included here. (See Figure 3.5.) Cemeteries, often city-owned, are important components of the urban forest. Sometimes dating back to early settlements, these grounds often support extremely diverse populations of trees, having been planted as living memorials to those interred there. From a management perspective, cemeteries offer excellent

FIGURE 3.4. Grounds of public buildings contribute to the urban forest landscape.

examples of where the benefits and liabilities of trees must be understood in context with the primary use of the land. (See Figure 3.6.)

Boulevard Medians. Medians in boulevards, separating opposing lanes of traffic, are commonly planted to trees and other vegetation. Vegetation may be for a variety of purposes: to add beauty, to influence traffic flow, to screen vehicle headlights, or to provide crash protection. (See Figure 3.7.)

Streetsides. Streetside tree situations vary greatly within individual cities. Trees are commonly planted on rights of way or strips adjacent to streets. These strips are known variously as treelawns, parkways, or parking strips. Such areas may be wholly city property or may be easements on private property. Rights and responsibilities concerning trees on easements are defined by ordinance. Varying widely by city, and often within cities, ordinances may give adjacent property owners total or partial responsibility or may grant rights to trees sufficient to claim injury damage. Total management responsibility by adjacent property owners is the exception, however, and in most cities the city forestry department is heavily involved with streetside trees. A national study found that an average of 61% of city forestry program budgets applied to street trees (Kielbaso et al., 1988). Four common streetside situations follow.

FIGURE 3.5. Historical monument grounds often have specimen trees.

Treelawn between curb and sidewalk (Figure 3.8). Most common in
   older residential neighborhoods, this situation is also found in
   modern subdivisions in the vicinity of schools. Widths of treelawns vary.
*Sidewalk immediately adjacent to the curb, with treelawn beyond,
   coalescing with adjacent yards* (Figure 3.9).
Sidewalk covering entire area between curb and building fronts (Figure 3.10). This situation is common in business areas. Trees, if
   present, are located in sidewalk holes or planters.
Streets with no sidewalks and no physically defined treelawns (Figure
   3.11).

FIGURE 3.6. Cemeteries often support a variety of tree species.

FIGURE 3.7. Trees in boulevards and medians give a sense of traffic separation and reduce glare from oncoming headlights.

FIGURE 3.8. Streetsides with treelawns between curbs and sidewalks are common in older neighborhoods.

FIGURE 3.9. When sidewalks are immediately adjacent to curbs, treelawns coalesce with adjacent yards.

FIGURE 3.10. In central business areas where paving abounds, trees are often located in special planters.

FIGURE 3.11. Streets with no sidewalks and no physically defined treelawns are common in modern subdivisions.

Streetsides, obviously, are in all parts of the city, in all residential, commercial, and industrial areas. They include new subdivisions, older residential areas, shopping centers, commercial strip developments, central business districts, and other areas of industry and commerce. They are endemic to areas of both opulence and poverty, and they serve all society.

Although streetsides often have areas called treelawns, they are rarely dedicated entirely to trees. They may also be used for both overhead and underground utilities: electricity, telephone, and television wires and cables; water mains and laterals; sewer pipes; and gas lines. Placement of utility lines varies greatly, depending on site and economic factors. In many cities, it is common practice to place water and sanitary sewer mains beneath streets, with laterals to adjacent buildings either beneath driveways or crossing treelawns. The modern trend is to place all utilities underground. (See Figure 3.12.)

Perhaps in no other area of the urban forest is compatibility of trees with other needs of society more important. Trees simply cannot interfere unduly with utilities, nor can they be undue impediments to vehicular and pedestrian passage and safety. Conversely, we must think about trees in terms of design, location, and maintenance of utilities.

We must also consider landscape design. (See Figure 3.13.) "A successfully designed streetside landscape will be open where pleasant views or safe vision is desired, closed where visual screen is needed; and varied in form, size, texture, and color for interest" (Nighswonger, 1986).

Thus, streetsides are areas where direct management is generally applied in total or in part by city forestry departments, in consideration of landscape design and spatial needs for other uses. As noted, streetside tree management also constitutes a major part of total urban forestry programs in many cities.

Parking Lots.   City-owned parking lots for employees, shoppers, commuters, sporting event patrons, and those attending cultural, civic, and other activities are often planted to trees, as are private parking lots. (See Figure 3.14.) Strategically located to minimize loss of parking spaces, trees provide needed shade for vehicles and are also important components in often otherwise harsh environments. That trees are valued in parking lots can be attested by early parkers who on summer days nearly always select spaces near trees. Parking lot trees offer special challenges. City officials who are concerned that parking spaces will be lost must first be convinced otherwise. Planting islands often must be constructed in areas of compacted soil. Reflective or heat-absorbing surfaces abound. Also, trees must be pruned to accommodate vehicles yet provide maximum shade.

FIGURE 3.12.   Streetsides are rarely dedicated to trees but are also easements for various utilities.

Riparian Areas. Riparian forests border streams, rivers, and lakes and occasionally include emergency flood drainage areas. Usually natural forests, these areas absorb storm runoff, provide valuable wildlife habitat, and are often used for picnic sites and hiking or nature trails. (See Figure 3.15.) Management of riparian areas may be designed to favor wildlife habitat, provide water access, or accommodate hiking, picnicking, fishing, or other day uses. Special attention must also be given, in high use areas, to tree hazard management. In many cities, riparian areas are separate linear parks, or extensions of conventional parks. A classic example is Rock Creek Park, extending from the Potomac River bordering Washington, D.C., into Montgomery County, Maryland.

FIGURE 3.13. Design of streetside landscapes is an important component of urban forest management.

Special Areas. This miscellaneous category includes situations that do not fit neatly as parks or other areas. Included are outdoor malls, fairgrounds, and markets. These areas are generally full of people and as a result have special tree management needs. (See Figure 3.16.) Trees are extremely important for aesthetic and environmental reasons and hence must receive high priority for management.

The preceding situations are the most common for which city forestry departments have direct management responsibility. Management is carried out by city crews or by contract with arboricultural, landscape, or nursery firms. For successful comprehensive management, all applications must be in context with the needs of the total urban forest.

FIGURE 3.14. Parking lots with trees are becoming increasingly common in the urban landscape.

FIGURE 3.15. Riparian areas are important elements of the urban forest.

FIGURE 3.16.   Trees enhance the environment of such special areas as outdoor malls and plazas.

Indirectly Manageable Urban Forest Lands

City forestry departments have indirect management responsibility for utility easements and other public or private lands.

Utility Easements. Utility easements are rights granted by landowners for construction and maintenance of utility services for the common good. In practice, easements are considered as physical spaces and include streetside and other strips on public and private property. Maintenance of utilities includes management of trees and other vegetation in the interest of uninterrupted services and public safety and is the responsibility of utility providers. From the standpoint of comprehensive urban forest management, city forestry officials must know the types, conditions, and extent of utility easements; who is responsible for them; and the management practices applied to each. Utility management is discussed further in Chapter 6.

Other Ownerships. City government has direct authority and responsibility, as discussed, for trees and other vegetation on city-owned lands and certain easements. Relatively little authority exists, however, on all other areas of the urban forest. Authority of city governments for urban forest management on private and other public properties is generally limited to situations of public health and safety and endangerment to the forest resource, such as insect or disease epidemics. It is on these myriad ownerships, both public and private, where indirect management must be applied. A description of each ownership within a given city is impractical, but such lands can generally be grouped as public or private, with various categories of each.

Public Lands—Federal and State. Federal government ownerships within the urban forest include office buildings, hospitals, monuments, cemeteries, and parks. State lands include highway rights of way, office buildings, universities, hospitals, and other institutions. Urban forest management on federal and state lands may be by in-service employees or by contract.

Private Lands. The privately owned urban forest may be best examined by the existing land uses, which include residential, commercial, industrial, undeveloped, and special use areas.

Residential areas vary widely by type, location, and economic conditions, from affluent neighborhoods in modern subdivisions to subsidized housing in high-rise buildings. Trees and related organisms generally receive a great amount of attention in residential areas. To many residents, the trees in their own yards are the most important trees in the entire urban forest. Depending largely on land use prior to development, urban forests of residential areas may be made up entirely of planted trees and other plants, or they may be mainly native trees of residual natural forests. The following factors influence the makeup and condition of residential urban forests.

Site planning and design, including location and specifications of streets, sidewalks, storm and sanitary sewers, other drainage facilities, utility lines, and other services.

Subdivision regulations, such as building setback and other spatial requirements, common area land percentages, and tree cover requirements.

Architectural considerations, including shape and size of buildings, solar exposure, and vistas.

Development and construction practices. Largely reflecting attitudes and economic interests of developers and builders and influenced by local regulations, trees and other natural vegetation may have been either protected or destroyed. In open areas lacking trees, developers may have planted trees and shrubs on individual lots and along streets before selling property.

Age of individual residential developments, which influences the size, composition, and condition of related urban forests.

In addition to these factors, all the elements that influence human choice—economics, attitudes, traditions, peer pressure, advertising, and marketing—also influence the residential urban forest. A few generalizations concerning economic influences may be made.

Greater priority is given to tree selection, planting, and maintenance in areas of higher income and in subdivisions and other developments having tree protection and management ordinances. Higher-income residential developments are also more often found in naturally wooded areas. (See Figure 3.17.)

Lower-income residential areas tend to be in older sections of cities, often interspersed with other commercial or industrial land uses. Tree populations are generally older, and replacement plantings are often not made when removals are necessary. When new plantings are made, fast-growing species are often selected.

Vegetation management on individual lots, particularly those with single-family residences, is the responsibility of owners and is done either personally or by contract with landscape and arboricultural firms. In other situations such as cluster or townhouse developments, management may be administered by homeowners' associations or similar organizations.

Urban forests of residential areas are not confined to streetsides and

FIGURE 3.17. Higher-income residential areas are more commonly located in natural forests.

yards, but often include natural areas, riparian zones, church grounds, cemeteries, playgrounds, parks, golf courses, and other recreational sites. Management of these areas may be by owners' associations or special boards. (See Figure 3.18.)

Commercial areas include central business districts, shopping malls (indoor and outdoor), plazas, strip developments, renovated waterfronts, historical districts, and other developments. Trees and other vegetation are vital aesthetic and environmental components of such areas and often must be established and maintained at relatively high cost because of site restrictions and people impact. Strip developments pose particular problems in that there are simply few places to plant trees to avoid overhead and underground obstructions. (See Figure 3.19.) Vegetation management, exclusive of streetsides and other areas of city responsibility, may be by individual business owners, corporations, or business associations.

Industrial lands vary widely but are generally classified as either light or heavy industry, ranging, for example, from companies producing computer chips to those manufacturing steel products. Generally, in areas of heavy industry, trees and other vegetation receive low priority. Little landscape planting is done, and if trees are present, they are often of volunteer origin. Light industries, however, are often in

FIGURE 3.18. Urban forests of residential areas often include natural woodlands, riparian zones, churchyards, cemeteries, parks, and golf courses.

FIGURE 3.19. Although trees could contribute greatly to commercial strip developments, there are few places to locate them.

planned industrial parks where streetsides, lawns, and other areas are carefully landscaped. Such areas may also be found in natural forests. (See Figure 3.20.)

Local governments or industrial development corporations own industrial parks and make strong efforts to attract new industries. Lands are then either leased, sold, or granted to new residents. Industrial parks normally involve large acreages with facilities for highway, rail, and even water transportation. Many airports also have adjacent industrial areas, serving air cargo and other needs. Landscaping of such areas, particularly streetsides and entrances, is often done to make them aesthetically pleasing to potential industries. Many industries also have a strong environmental sense and develop attractive landscapes for employees and visitors. Responsibility for urban forest and landscape management is normally with corporate residents. On unused lands, though, local government or other owners are responsible. Management is most often by contract, administered by an individual or department within each corporation. Occasionally, such administrators have formal training in urban forestry or landscape management.

Undeveloped lands are found in nearly all cities. These lands may be parcels awaiting development, abandoned industrial sites, lands held

FIGURE 3.20. Trees and other landscape plants can greatly improve the environment of industrial areas.

in estates, or areas where topographic or other physical restrictions make development impractical. They are most often in a natural state with abundant vegetation. Many such areas, particularly in eastern cities, support mature native forests. For the most part, these lands are virtually unmanaged, receiving scant attention except for fire suppression. Occasionally, however, access rights are obtained for hiking or nature trails and other public uses. Also, some tracts produce fuelwood and other forest products. Values of such lands are considerable: wildlife habitat, soil erosion control, screening, sound attenuation, and aesthetics. Screening is of particular value, as buildings and other structures are often hidden, giving the impression of low population density. (See Figure 3.21.) Management, as may be practiced, is the responsibility of owners or holders of special rights.

THE INSTITUTIONAL AND COMMERCIAL ENVIRONMENT

For comprehensive urban forest management to be effective the city forestry department must develop a relationship with other units of city

FIGURE 3.21. The urban forest often includes areas of mature natural forests.

government, those who provide labor and services, and others who influence the urban forest.

Other Units of Government

Units of government other than city forestry departments also influence urban forestry. An example is a street department whose construction and maintenance work directly impacts trees. The activities of nearly all other departments—parks, recreation, water, refuse, fire, police, planning, and others—often concern the urban forest. Conversely, city forestry departments must treat trees in conformance with standards for which other departments are responsible. Such standards may be prescribed by city code. Tucker (1994) found, for example, 19 references to trees and other vegetation pertaining to other departments in the city codes of Westminister, Maryland.

   Trees can be impediments and may be seen only as such, particularly by engineering, construction, and maintenance individuals. It is important that requirements and concerns of these individuals be understood and that they be made aware of techniques and technologies to make trees compatible with the structures for which they are responsible.

Providers of Labor and Services

Essential to comprehensive management is knowledge of those who provide services and physically work in the urban forest. Such individuals and firms leave their mark on the urban forest to a greater degree than perhaps anyone else, for they physically plant, maintain, and remove trees and other plants. It has been correctly said that no one influences the forest more than the person with a tool in his or her hand. It follows that one of the highest priorities of comprehensive management should be to improve the knowledge and skills of those who hold the tools. This is discussed in greater detail in Chapter 7. A listing of those who provide services and labor follows:

Arborists—commercial tree care firms and institutional employees
Landscape architects and designers
Groundskeepers—individuals employed by institutions or companies or commercial firms providing contract services
Nurseries and plant material retailers—production nurseries, retail nurseries, and other plant material outlets

Planners, Designers, and Developers

Growth, development, and redevelopment of cities have profound impacts on the urban forest. The process, involving planning, design, and construction, is complicated. Planning services are provided by the city planning department and by commercial firms. Design by landscape architects and other professionals determines to a great degree the face of the urban forest. Construction activities are often detrimental to trees and other vegetation, but damage can be minimized with planning, education, and supervision. Working with planners and developers is discussed in Chapter 7.

Other Players

Many individuals and organizations either are directly involved with the urban forest or may exert a strong influence. These players include government agencies, businesses, and citizens' organizations.

Other Government Agencies. Two government agencies (exclusive of city government) strongly influence the urban forest: the Cooperative Extension Service and the state forestry agency. The Cooperative Extension Service has offices in nearly all counties in the nation. Extension agents dispense a wide variety of information related to the urban forest and provide such services as soil testing, meeting coordination, and demonstrations. They also make available university specialists in plant health management, landscaping, plant physiology, and other disciplines. Many agents write newspaper columns relating to horticulture and forestry, and some host radio and television programs. State forestry agencies have a mandated responsibility for urban forestry: working with city governments to establish and administer programs, sponsoring training in urban forestry and arboriculture, and administering grants for special projects such as landscaping, tree inventories, and educational materials.

Business and Citizens' Organizations. In every city there are public and private institutions with which the city forester may work to effect comprehensive management. Each contact provides opportunities to communicate information, obtain funds and services, and work with volunteers. These institutions include garden clubs, wildlife clubs, beautification committees, homeowner and neighborhood associations, youth organizations, business organizations, civic and service clubs, churches, and the media. The city forester's relationship with these institutions, with particular emphasis on working with volunteers and the media, is discussed in Chapter 8.

THE LEGAL ENVIRONMENT

The legal environment in which urban forestry operates has three primary components: ordinances, regulations, and liability considerations. This classification does not ignore the total legal environment affecting all society, but it is specific to legal matters concerning urban forestry operations.

Ordinances

Urban forestry ordinances are city or local government codes concerning trees and related organisms. Ordinances are of two general types: those providing for management and those concerning tree protection and landscape matters. Within these types, Bernhardt and Swiecki (1991) identify three categories: street tree ordinances, tree protection ordinances, and view ordinances. The following discussion involves these categories.

Management Ordinances. Ordinances providing for management give authority, define responsibility, and set forth minimum standards for safety and convenience. The effect is to charter city forestry programs and provide a legal structure for operations. Management ordinances generally have eleven basic elements:

1. Definitions
2. Designation of city forester (general responsibilities)

3. Establishment of a tree board (composition, terms of office, duties, procedures)
4. Responsibility for trees on city property and easements
5. Planting regulations (permits, official species, spacing, location)
6. Maintenance standards
7. Removal requirements and standards
8. Catastrophic authority on private property (condemnation, treatment)
9. Requirements for private contractors
10. Prohibition of interference
11. Violations and penalties

All these elements may not be in a single document, particularly items 5–7, which are often in a separate "standards" section. Such standards technically may not have the weight of law and may more properly be considered as policies for operation. It is important to recognize, as noted previously, that legal requirements concerning trees may also be found in ordinances concerning other departments of city government. Management ordinances are often not comprehensive, many having evolved over time, with new sections added as situations arose. The origins of ordinances is not important. What is important for comprehensive management is that ordinances provide a solid legal framework for professional operations.

Tree Protection and Landscape Ordinances. Tree protection and landscape ordinances relate generally to special trees and groves, development and construction, and special landscape needs, including views and solar access. Responsibilities of urban forest managers in these situations may include identification and classification of special trees, review of site and landscape plans, and design and enforcement of specifications for tree protection during development.

Special trees and unique groves are often protected by ordinance. Protection may be because of size, species, scarcity, historical or cultural significance, position in the landscape, or environmental contribution. Often referred to as tree preservation ordinances, such laws establish criterion for special trees and groves, allow them to be set aside, and authorize maintenance. (See Figure 3.22.)

Ordinances relating to development and construction concern the maintenance of tree cover and the protection of trees. Tree cover requirements vary and may be expressed as a percent of total land area, stems per unit of land area, or crown cover. Critical drainage areas, wetlands, and other special ecosystems may also be included. Protection of individual trees during development and construction is difficult to define by ordinance. Hence, wording tends to be general,

FIGURE 3.22. Special trees, particularly those of historical or cultural significance, often are protected by ordinance.

requiring that due care be exercised to prevent damage. Standards or operating procedures for builders and contractors may be prescribed, but such cases are exceptional.

Ordinances concerning special landscape matters are generally of two types: those that require tree planting and other landscaping for new developments and those that relate to diminishment of economic value, solar access, enjoyment of views, and general beneficial use because of location and establishment, growth, or maintenance of trees.

Regulations

Regulations concerning urban forestry have the weight of law and might properly be considered as ordinances. They are considered separately, however, because of their purposes and origins. Regulations may be classified as those concerning homeowners such as subdivision regulations, those concerning program administration, and those based on state or federal legislation.

Subdivision Regulations. Most modern subdivisions have regulations for the "common good." Many properties also have deed restrictions that influence trees and other vegetation. Based on concerns for maintenance of property values, public safety, environment, and other factors, subdivision regulations encumber developers and subsequent property owners. Regulations concerning vegetation may include species selection, planting location, insect and disease control, and maintenance. Subdivisions commonly have homeowners associations, with approval and enforcement authority vested in a board of directors or landscape committee.

Administrative Regulations. Administrative regulations are requirements of city government that facilitate or influence urban forestry operations. Included are registration and licensing requirements for arborists and other landscape operators, business permits, and regulations concerning use and disposal of wood and other vegetative materials.

State and Federal Legislation. Often reflected in city ordinances, state and federal laws have a strong influence on urban forestry. Such laws concern both the natural and work environment and involve

Rare, threatened, or endangered species
Wetlands or other critical habitats
Point and nonpoint source pollution
Pesticides
Workplace safety
Workers' rights

Liability

Trees often pose threats to life and property. Entire trees may topple, branches may fall, leaves and fruit may create slippery surfaces, foliage and trunks may obscure views, and roots may heave sidewalks. Resulting accidents often leave property owners and those responsible for tree care liable for damages. The trend toward increasing litigation in modern society, plus increasingly higher damage awards by courts, makes the issue of liability extremely important in urban forestry. Liability is based on the tort law principle of prudent and reasonable care. Thus, property owners and their agents have a responsibility to exercise such care. To city forestry departments this responsibility translates into effective and on-going tree hazard management programs. Tree hazard management is discussed in Chapter 6.

THE POLITICAL ENVIRONMENT

The influence of politics is a reality. In cities with highly successful urban forestry programs, it cannot be ignored. It is also a reality that within cities there is intensive competition for funds; those programs and services having key political support are generally well funded. Political support is a manifestation of citizen concern, reflecting the wishes of the general public or of key individuals who influence city administrators. Thus, it behooves the city forester to learn the political environment and understand the forces influencing city government. Such involvement in the political environment may be distasteful to some, but to ignore it is to ignore reality.

SUMMARY

The importance of understanding the urban forestry environment cannot be overstated. The physical environment involves the urban forest itself—where it is located, the purposes it serves, who tends it, and its management constraints and opportunities. Within the institutional environment are the societal resources for management—sources of funding, services, technical assistance, volunteers, and opportunities for communication. The legal environment includes the framework for management and mandates certain practices. The political environment is a reality and must be understood to the advantage of urban forestry.

Included in the foregoing discussion is the concept of three categories of managers of the urban forest: city government, public utilities, and individual owners. Within these groupings are all who make decisions about what is to be done to the urban forest. These categories represent three general audiences for urban forest managers who are attempting to improve the total urban forest.

REFERENCES

Bernhardt, E. A., and T. J. Swiecki. 1991. *Guidelines for Developing and Evaluating Tree Ordinances,* California Department of Forestry and Fire Protection.

Kielbaso, J. J., et al. 1988. Trends in urban forestry management. *ICMA Baseline Rep.,* 20(1): 12.

Nighswonger, J. J. 1986. "Management of the Urban Forest—Design," *Urban Forestry,* John Wiley & Sons, Inc., New York, p. 150.

Tucker, K., City Planner, City of Westminister, Maryland. November 1994. Personal communication.

CHAPTER 4

Determining What the
Urban Forest Needs

You gotta know what the urban forest needs.

W hat does the urban forest need? The answers to this question are fundamental to setting priorities and directing efforts to where they will do the most good. What must first be determined is the present condition of the urban forest, in all parts of the urban landscape. Only with this information, can judgments of need and subsequent recommendations for management be accurately made. Nowak (1994) referred to the need for structural information:

> Management of the entire urban forest ecosystem requires information on all vegetation and other attributes of the system across the urban landscape. This type of structural information establishes a basis for comprehensive management that recognizes linkages among the multiple land uses and owners of the urban forest. Forest structure also provides a means to estimate the actual and potential physical, biological, social, and economic functions of the urban forest. Urban foresters can then develop plans and programs that provide for these functions across the urban landscape.

Some elements of the present condition of the urban forest include health, vigor, safety, diversity, stocking, functionality, and aesthetics.

35

These elements must be translated into quantifiable needs for the future—numbers of trees to be established, pruned, removed, or otherwise treated and measures for ensuring continuing safety, diversity, and functionality. Determining such would be relatively easy if the urban forest were homogeneous, rather than the complex mixture discussed in Chapter 3. Needs must be determined for the various parts of the urban forest, and priorities must be assigned in context with the whole.

We can determine what the urban forest needs in several ways. We can simply observe while riding or walking and get a general idea of whether a particular area is in need of additional planting, hazard removal, corrective pruning, or other action. We can also gain information as to relative sizes of trees from low-altitude aerial photographs. Infrared and other photographs can reveal certain insect or disease infestations or other abnormalities. Also, city records provide development dates for subdivisions and other areas. From such information we can draw general conclusions regarding management needs. We can also use work records of city forestry departments and utility companies to determine when pruning or other work was last done. An additional way is to wait for complaints. Unfortunately, this method is used in some cities, and management is by crisis rather than by plan. Precise information, however, must be obtained by surveys and inventories.

To determine overall need, we should begin with cover-type mapping of the entire urban forest to identify general vegetative communities. Cover-type mapping should have as its base the various land use situations identified in the Physical Environment section of Chapter 3. We can generally locate such situations from standard maps and aerial photographs. We can then use aerial photographs to identify open areas such as grass, tilled fields, wetlands, open water, and tree-covered areas by type (hardwoods or conifers), size, density, and arrangement. Ground checks can confirm or further refine type identification. The product is a general vegetative cover-type map of the entire urban area by land use situations. The next step is to superimpose on the map those areas that may be managed directly and those that must be managed indirectly. Where practical, this process may be speeded and greatly refined by the use of Geographic Information Systems (GIS). Using advanced software, we can use computers to scan visual images from aerial photographs, maps, and even satellites, allowing features to be delineated as layers. Coupled with ground checks and other information, fully detailed map images may be developed.

Perhaps nowhere is the concept of direct and indirect management of the urban forest more important than in determining needs. In general, detailed information is needed for areas that are to be managed directly.

Less precise (but accurate) information is needed for indirectly managed areas. From the standpoint of the city forester, there are three factors involved in determining the degree of information needed: (1) responsibility for management; (2) degree of management as determined by function and environment of individual trees; and (3) overriding circumstances. The first factor is a restatement of direct and indirect management but involves the stage of city forestry program development. The second factor may be explained by the example of trees in high public use areas requiring intensive care, as compared with those in natural areas, which require only extensive management. The third factor, overriding circumstances, is something that may be common to the entire urban forest, such as storm damage or a particular insect or disease epidemic. These three factors must be weighed by the city forester when he or she is determining the type and amount of information to be obtained.

Urban forest information needs are somewhat different for a functioning city forestry program than for a program in early states of development. With the former, the need is for site specific information to facilitate establishment, maintenance, and removal. This information must be obtained through formal inventories. With the latter, the need is for overview information to develop a long-range plan or to help make a convincing case for putting in place or improving a city forestry program. This information can be gathered by less formal surveys. Surveys will be considered first, as they are preliminary to full program development.

SURVEYS

The need is for relatively low-cost information to facilitate long-range planning or program development. Thus, we must obtain different information for lands that require direct management than for those lands that require indirect management. For areas for which city is responsible and direct management is to be applied (streetsides, parks, other areas), we need to know

Approximate number of trees
Species
Average size and age
Condition
Stocking

For areas for which the city is not responsible and indirect management is to be applied (private and other public lands), we need to know

Vegetation types
Condition

Overriding circumstances

From this information, we can quantify future direct management needs: number of trees to be planted, amount of pruning needed, number of trees to be removed, and other practices. Reasonable conclusions can also be drawn as to indirect management needs for species and age diversity, health protection, and improved arboricultural practices.

The Survey Process

The survey process involves identifying and mapping land use situations as previously described; vegetative cover mapping within each situation; ground observations and point sampling, if needed, of private and other public lands; and formal sampling of areas for which the city has management responsibility.

Areas for which the city has responsibility require a major part of the survey effort. These areas are generally streetsides and parks. Streetsides by homogenous areas such as subdivisions, neighborhoods, or other commonalities may be surveyed from a slowly moving vehicle. Trees are tallied by species, estimated diameter, and condition classes reflecting management needs. Empty planting spaces are also tallied, as are stumps to be removed. Examples of condition classes follow.

Good. Healthy, vigorous tree. No apparent signs of disease or mechanical injury. Little or no corrective work needed. Form representative of species.

Fair. Average condition and vigor for area. May be in need of some corrective pruning or repair. May lack desirable form characteristic of species. May show minor insect injury, disease, or physiological problem.

Poor. General state of decline. May show severe mechanical, insect, or disease damage, but death not imminent. May require major repair or renovation.

Dead or dying. Dead, or death imminent from mechanical, disease, or other causes. Removal needed.

Survey samples vary as a result of the size of the area to be surveyed and time and money constraints. Total street miles must be determined, and representative streets need to be surveyed. A common practice is to survey every third or fourth street. Table 4.1 is a summary from a streetside tree survey of a section of a small midwestern city (58 street miles). The summary reveals that there are

High percentages of American elm and hackberry
High percentages of larger, older trees

TABLE 4.1. *Sample Streetside Tree Survey Summary*

Species	Number of Trees	Average Age	Average Diameter (in.)	Total %				Percentage of Total Trees
				Good[a]	Fair[b]	Poor[c]	Dead or Dying[d]	
American elm	2,158	60	19	31	25	24	20	33
Hackberry	1,090	60	16	59	31	10	0	18
Siberian elm	376	40	14	4	33	59	4	6
Hard maple (sp.)	328	10	4	74	5	18	3	5
Green ash	320	15	6	48	31	21	0	5
Pin oak	320	40	15	69	10	20	1	5
Sycamore/planetree	302	20	8	76	9	15	0	5
Honeylocust	288	20	9	71	8	20	2	4
Silver maple	280	15	10	39	13	48	0	4
Red oak	218	10	5	56	23	13	8	3
Redbud	138	10	4	78	2	20	0	2
Black walnut	130	55	15	34	54	12	0	2
Hybrid elm	82	15	6	7	56	37	0	1
Miscellaneous[e]	466							7
Totals	6,496							

[a]Healthy, vigorous tree. No apparent signs of insect, disease, or mechanical injury. Little or no corrective work required. Form representative of species.

[b]Average condition and vigor for area. May be in need of some corrective pruning or repair. May lack desirable form characteristic of species. May show minor insect injury, disease, or physiological problem.

[c]General state of decline. May show severe mechanical, insect, or disease damage, but death not imminent. May require repair or renovation.

[d]Dead, or death imminent from Dutch elm disease or other causes.

[e]Less than 1% each: bur oak, tree of heaven, flowering crab, sweetgum, basswood, black locust, Kentucky coffeetree, eastern redcedar, chinquapin oak, hawthorn, mimosa, catalpa, pine (sp.), cottonwood, willow, Osage orange, purpleleaf plum, white birch, mulberry, fruit (sp.), Japanese pagodatree, shingle oak, English oak, Russian olive, ginkgo, boxelder.

A large number of Dead or Dying America elms (430 trees)

Generally high percentages of Good and Fair trees, with the exception of Siberian elm and silver maple

A large number of adaptable species

Overrepresentation of American elm and hackberry

Relatively low stocking—56% based on an optimum stocking rate of 200 trees per street mile

From this information, the following general conclusions as to future management are reasonable.

1. Removal of dead and dying trees is necessary and must receive high priority for safety reasons.
2. Control measures are needed to protect the remaining American elm population.
3. A major planting effort (5,104 trees) is needed to achieve optimum stocking and species diversity.
4. A pruning effort is needed to correct and restore Poor and Fair trees.

Parks and other areas may be surveyed by using the same criteria of species, diameter, and condition. Transects or ground grids can be used to avoid either missing or double-counting trees.

It is important to note that in surveys of streetside and park trees identifying locations of individual trees is generally not necessary. Thus, information is collected by areas within land use situations. The need is to present a picture of the total urban forest, with accurate conclusions as to overall management needs.

MANAGEMENT INVENTORIES

Management inventories differ from surveys in that they locate individual trees, are generally more detailed, and may be continuous. They are also more expensive. Most commonly applicable to streetsides, management inventories are directed ultimately to individual trees. Because management inventories are very expensive, the information collected for each tree must be directly applicable to future management. Collecting more information than is needed is costly. Collecting less may be equally costly. Hence, we must plan management inventories carefully, starting with the basic question: What must we know about individual trees in order to best manage them in relation to their environment? Keeping with basics, we need to know three things: what the tree is, what it needs, and where it is. As with most basics, elaboration is needed.

To know what the tree is, we must know its species, size, and age. Size, as expressed by diameter and crown spread, is an indicator of age. Thus, a record of a red oak tree, 40 inches in diameter, with a crown spread of 70 feet creates a mental image of "a big old tree" and suggests to the experienced urban forester that it may have special problems.

Knowing what it needs is the bottom-line answer to what should be done to the tree and/or its immediate surroundings to optimize the tree's role in the urban environment. We make this judgment according to the characteristics of the tree and its surroundings. It may be made on the spot by the person conducting the inventory or later according to information collected. The advantage of a spot judgment is that less time and recording expense is required. The disadvantage is that it is a need recorded for a single point in time and does not provide a record of the factors responsible for the need. Thus, most inventories include indicators of tree condition such as presence or absence of dead or broken branches; evidence of insect, disease, or physiological problems; and factors of the tree's growing environment.

Knowing where it is is critical to management. Work crews simply must be able to locate individual trees precisely. Urban forestry is replete with stories of the wrong tree being treated or even removed. Locating streetside trees presents special problems. Addresses can be confusing and lot boundaries are often difficult to identify on the ground. Obtaining precise distances from street corners or other "bases" is time consuming and costly. Sequential numbering from street corners is invalidated by new plantings. Of the above, lot numbers are most commonly used; and with accurate records of species, size, and other characteristics, locating individual trees is possible.

A summary of information needed for individual trees follows:

I. What it is
    A. Species
    B. Diameter (4½ feet from ground)
    C. Crown spread
II. What it needs
    A. Treatment needed
        1. Pruning, by type
        2. Cabling, bracing, other mechanical
        3. Insect or disease treatment

        4. Soil treatment—fertilization, aeration, other
        5. Removal of physical impediment
        6. Tree removal
    B. Indicators of need
        1. Trunk condition
           a. Normal for species
           b. Forked
           c. Wounds
           d. Evidence of decay
           e. Other
        2. Branch condition
           a. Normal for species
           b. Broken
           c. Dead
           d. Obstructing
           e. Other
        3. Crown and leaf condition
           a. Normal for species
           b. Crown dieback
           c. Leaf discoloration or abnormality
           d. Stem cankers or other conditions
        4. Environmental factors
           a. Width of treelawn
           b. Proximity of overhead wires
           c. Underground utilities
           d. Apparent soil problems
           e. Rootzone restrictions
           f. Obstructions
           g. Other
III. Where it is
    A. Street name
    B. Lot number or other location method

With the exception of location factors and information about underground utilities, all data must be collected on site, requiring trained personnel and data processing systems. In most cases, data processing is done by computer. Data entry is sometimes done manually, but the trend is toward automatic entry systems from the field. Computers provide for multiple cross-tabulations and allow near-instant information retrieval. The more sophisticated systems can visually display individual trees or streetside situations. For the urban forest manager, this means instant access to the condition and need of every tree within his or her streetside jurisdiction.

Inventories may also be for special purposes, such as land condemnation, wildlife habitat management, species or ecosystem preservation, or tree hazard evaluation. In such cases, objectives must be clearly defined, and systems must be designed to produce consistent outcomes.

Inventories are snapshots in time and, without provisions for continuing updating, quickly become obsolete. For continuous inventories a system for information feedback and an allowance for tree growth is necessary. Information feedback systems must ensure that everything that changes a tree—prescribed treatments, construction, accidents, storm damage, other—is recorded. A common method for recording prescribed treatments is by returned work orders. Changes resulting from construction, accidents, storms, and other causes, however, are more difficult to identify and record. Growth changes are likewise difficult to predict, requiring judgments based on sampling and observation. Site variability presents an obvious problem when attempting to determine average or normal growth rates by species. Reasonable judgments can be made, however, and unless changes are monitored and accurately recorded, inventories quickly become outdated, with lifespans rarely exceeding 5 to 7 years.

Inventory information is largely the basis for all planning and subsequent operations. Consequently, it is of obvious internal value. Inventory information has external value also, as it reveals situations of which property owners, arborists, nursery people, and others should be aware. A common example is the lack of desirable species diversity. If nursery people and others are made aware of this problem, other species can be made available, promoted, and established in the urban forest.

Much has been written about urban forest inventories, and many systems have been developed. A bibliography concerning urban forest inventories appears at the end of the book.

SUMMARY

Fundamental to comprehensive management of the urban forest is an accurate assessment of needs. What must first be determined is the present condition of the urban forest, including its health, vigor, safety, diversity, stocking, functionality, and aesthetics. Only with this information can plans be developed, resources allocated, and management applied. In general, more detailed information is needed for areas that are to be managed directly than for areas that are to be managed indirectly. The degree of information needed is further influenced by the function and location of individual trees (which determines the intensity of management needed) and the stage of city forestry program development. It is critically important that the degree of information be determined by the need. The condition of the urban forest can be assessed by surveys or inventories. Surveys provide overview information and are generally applicable to long-range planning. Inventories are formal and provide detailed information (including specific location) about individual trees. Inventories are the basis for direct management, providing both baseline and continuing condition information.

REFERENCE

Nowak, D. J. 1994. Understanding the structure of Urban Forests, *Journal of Forestry,* 92(10): 42.

CHAPTER 5

Planning and Budgeting
for Urban Forestry*

You gotta have a plan.
You gotta have money.

P lans are necessary; therefore, planning is necessary. The urban forest changes; therefore, plans must change. These simple statements confirm the absolute necessity of urban forestry program planning. Planning:

Ensures that all needs of the urban forest are recognized
Reduces the risk to property and human safety
Makes sure that all work is prioritized
Provides a defensible basis for budgets and grant requests
Leads to a continuous program from year to year, regardless of personnel changes

Planning starts with the understanding that there must be a long-range plan and that operational plans, budgets, and plans of work follow. The long-range plan states the mission, defines objectives, identifies strategies, and determines priorities for achieving objectives. Operational plans are based on the long-range plan and are incremental

*This chapter is based on "Planning for Urban Forestry" from *A Handbook for Tree Members* by G. W. Grey and published by The National Arbor Day Foundation in cooperation with the National Association of State Foresters. Please note that the planning process outlined is basic, allowing for modification to meet needs of individual cities.

to achieving objectives. Budgets are based on operational plans. Plans of work are tasks and timetables for carrying out operational plans. The final aspect of planning is evaluation. Thus, the planning flow is as follows:

Long-range planning. States the mission and defines objectives. Identifies strategies, or means, to achieve objectives. Usually a function of tree boards and/or city forestry departments.

Operational planning. Incremental (usually annual). Based on long-range plan. Outlines strategies. Usually a function of city forestry departments.

Budgets. Based on items in operational plans.

Plans of work. Internal documents for city forestry departments and tree boards, giving tasks and timetables.

Feedback and evaluation. Ongoing, with emphasis on end-of-year review.

As suggested by the planning flow, planning need not be difficult. Yet, planning is feared by some and misunderstood by many. Planning has two basic criteria.

1. Plans must include
   a. An assessment of what you have
   b. A vision what you want it to be
   c. How you are going to get there
   d. What it will take to get you there
   e. Assessments of how you are doing
2. Plans must be based on accurate assessment of accurate information.

In urban forestry planning, the ultimate vision is the same for all cities: a forest that is safe, healthy, adequately stocked, diverse, pleasing to the senses, and functional. How the vision will be achieved requires strategies, priorities, and incremental goals. It also requires money, personnel, and other resources. The second basic criterion (accurate information) is an absolute essential. There must be adequate knowledge of what the urban forest needs—from surveys or inventories, as discussed in Chapter 4.

LONG-RANGE PLANNING

The long-range planning process has two objectives: a documented plan for the entire urban forest and acceptance of the plan by decision makers. A common objection to planning is that plans are nothing more than "shelf documents" that no one pays any attention to. Unfortunately, urban forestry plans are sometimes unused. The reasons are twofold: lack of acceptance by city administrators and lack of commitment by program managers. Thus, in addition to producing a plan, planning must embrace the desire to seek acceptance of and commitment to the plan.

The long-range planning process involves four steps:

1. Planning to plan
2. Planning
3. Review
4. Acceptance

Planning to Plan

The first consideration is who should be involved in planning. Leadership for planning is often mandated by ordinance or administrative directive. Planning is a commonly stated activity in ordinances governing tree boards. In other cases, plans may be directed by city commissions or initiated by city forestry departments. Tree boards offer a tremendous advantage in planning in that individual members represent the interests of citizens, the ultimate stakeholders, in the urban forest. Citizen interest is critically important, and public input must be sought during the planning process.

It is advisable to have two planning groups: a primary planning team and a planning advisory committee. The primary planning team will vary by city but should include those directly involved, such as tree board members, the city forester, perhaps another related department head such as the parks director, and preferably a city administrator. The responsibility of the primary group is to direct and conduct the planning process. The advisory committee's role is to ensure that the interests of citizens are represented. The advisory committee can also be a major key to acceptance of the plan. Committee members should be "direct" stakeholders, such as nursery or tree care representatives, volunteer group leaders, and land developers—including those who might benefit from the plan and those who may be threatened.

Once assembled, a first order of business of the primary planning team is to develop a planning plan and a timetable. The planning plan should include an outline of the planning process and a list of reference resources, including survey or inventory information, ordinances, and current program status reports. The timetable must be consistent with the planning outline and consider other commitments of the planning team.

Planning

As indicated previously, planning must start with a vision. Even though the vision may be generic, it is important that the planning team develop a statement for their urban forest. By so doing, it becomes their vision, understood and agreed upon by all, and is the focus of all that follows. The vision can be developed by simply listing the things the urban forest should be in the future—safe, healthy, well stocked, diverse, functional, aesthetically pleasing, and such. A vision statement can then be drafted by putting the list into a declarative paragraph as in the following example.

> As representatives of the citizens of our city, we who are responsible for urban forestry planning commit to achieving by the year 2015 an urban forest that is safe, healthy, well stocked, diverse, pleasing to the senses, and functional—an urban forest attractive to birds and other wildlife, and a place in which our children may delight and learn.

Although such a vision statement may be idealistic and even poetic, it expresses commitment and provides inspiration. The statement also includes program objectives, painlessly developed, yet clearly stated. There can be debate about whether setting a date for achieving objectives is advisable at this time. The date is perhaps helpful in that it indicates a planning horizon, but it should be considered as tentative pending a close look at resources relative to total needs.

The next step is to list strategies, or means, to achieve each objective identified in the mission statement. Strategies should be simple statements of what can be done. For example, for each objective identified in the preceeding vision statement, the following strategies might be listed.

To achieve a safe urban forest:

Identify and remove, where hazards exist, materials including dead trees, dead or weakened portions of trees, roots, obstructions caused by roots such as heaved sidewalks, fruits, and other materials produced by trees.

Remove trees or portions of trees interfering with pedestrian or vehicular passage or with traffic regulatory signs or signals.

Remove trees or portions of trees in contact with high-voltage lines or structures.

Remove targets in cases of unique hazard trees.

Prune trees regularly and use other arboricultural practices to ensure structural strength.

Plant potentially safe and vigorous trees.

To ensure health:

Implement a plan of plant health management including maintenance of vigor, monitoring of insect and disease situations, and
selective treatment.

Select and plant diverse tree species and varieties potentially free
from excessive insect, disease, or environmental damage.

Prune to remove dead and broken material and to help prevent future
damage.

To achieve adequate stocking:

Plant trees in vacant spots along streets and other selected areas.

To create diversity:

Plant a variety of species over time.

Remove decadent trees of overabundant species.

Landscape new areas with a variety of species.

To make the urban forest pleasing:

Incorporate landscape design considerations, particularly with regard to new plantings and other developments.

To benefit wildlife:

Plant trees and other vegetation favorable to birds and other wildlife.

Protect wildlife habitat during development and construction.

Retain dead trees and portions of trees for cavity nesting birds where
such retention does not present a hazard.

Erect artificial nesting structures.

An executive summary of a long-range plan, including the foregoing
objectives and strategies, is given in Appendix 2.

Once listed, the strategies must be quantified, calling on information
gathered from surveys, inventories, or other sources, in order to provide
answers such as the number of trees needed to achieve adequate
stocking; the number and sizes of hazard trees in need of removal,
pruning, or other treatment; the amount of pruning necessary to correct
or prevent interference with pedestrian or vehicular traffic; and the
amount of pruning needed to provide for future tree health and vigor.
When quantified with accurate information, a picture emerges of the
total effort necessary to set the urban forest in order. It must be
recognized, however, that the urban forest is a moving target, with
constantly changing needs. Thus there is a need for continuing inventories or other methods of updating information and the necessity of
revising plans.

At this point it is necessary to estimate costs of the various strategies. The long-range plan must have a bottom-line cost figure, because during presentation to city administrators, someone will certainly ask the question: How much is all this going to cost? And costs, reasonably estimated and well documented, give credibility to the plan. A total cost figure also allows annual, or incremental, program cost estimates to be made. Planting (establishment) costs are relatively easy to estimate. Quotes for on-site planting may be obtained, or wholesale prices can be requested and the rule of thumb of establishment costs, which equals two and one half times wholesale costs, applied. Pruning, removal, and other treatment costs are more difficult because need, tree sizes, and other factors vary widely. Sample estimates may be obtained, or recent examples of such work examined.

Priorities also must be identified. Without question, top priority must be given to making the urban forest safe. Safety is both a legal and moral responsibility of urban forest managers. Subsequent priorities must be based on assessment of needs as indicated by survey or inventory information. Priority assignment allows work to be projected over the period of years necessary to reach the objectives. Such projections must be based on reasonable estimates of what parts of the total job can be accomplished yearly with existing and potential resources.

Review

Review has two purposes: to get further input and to gain support. Primary reviewers are review committee members, although it may be desirable to "air" the proposed plan at public meetings or through the media. The review document should include:

- A statement of need for the plan, supported by current urban forestry situation information from surveys or inventories
- A draft plan, including objectives, strategies, and priorities
- Cost estimates
- Names, addresses, and telephone numbers of members of the planning team and review committee

Well before a meeting, the draft document should be furnished to each reviewer, with a cover letter giving directions and recommendations. Reviewers should be asked to consider specific things, such as other strategies that might be employed, additional resources that might be available, and additional things concerning the urban forest not addressed in the plan. Reviewers are thus given the opportunity to provide positive input rather than to serve only as critics, which hopefully will increase their feelings of ownership and the likelihood of their support.

Acceptance

While formats of plans may vary, the final document to be presented to city administrators should include the following elements:

- Current situation statement and statement of need for a long-range program
- Vision statement, including objectives
- Strategies to reach individual objectives
- Statement of priorities based on strategies
- Cost estimates
- Supporting information

Whether the plan is approved will depend on the quality of the plan, its reasonableness, and its support by citizens. Within the reasonableness factor are cost/benefit considerations. A convincing case must be made for the plan's ability to save money in the future—to prevent costly problems by properly selecting and locating new plantings, by scheduled pruning, and by paying prompt attention to hazardous trees. Such cost savings can be documented, particularly in the case of hazardous trees, in view of the city's liability responsibilities. As indicated previously, it is also extremely important to document and explain all cost estimates—to show that each is based on the best available information and judgment. In the final analysis, the plan must be judged by the majority to be in the best future interests of the city and its citizens.

OPERATIONAL PLANS AND BUDGETS

Operational plans carry the action. They include the details of how strategies identified in long-range plans are to be carried out. Operational plans are incremental, usually annual, and based on the city's fiscal year and budget cycle. At the outset of planning, two questions must be asked: What are the priorities? and How big a chunk of the total job can be handled next year? During the first few years hazard trees will probably receive the highest priority. While making the urban forest safe must always be a top priority, the proportion of time spent on hazard control should decrease over time.

Based on priorities, operational plans are basically statements of how each item will be accomplished and what is necessary for accomplishment. In view of the two questions identified in the preceding paragraph, planning is a matter of determining goals and resources:

Goals. Targets that are quantifiable by numbers of trees to be planted, numbers of dead trees to be removed, amount of pruning to be accomplished, and other treatments to be applied

Resources. Capital that is applied to the costs of plant materials, equipment (perhaps amortized), materials, and services, necessary to achieve goals

Goals and resources must be based on such decisions as species of trees and where they are to be planted, types of pruning and location of trees to be pruned, and whether work is to be done in-house, by contract, or by volunteers. Figure 5.1 is a planning and budgeting worksheet, showing how such considerations might be organized.

Budgets are simply documented statements of program costs and

---

Planning Year: _____
Planning Date: _____
Relevant Strategies from Long-Range Plan:_____

Goal: (1)_____

Resources Needed: (2)	*Estimated Costs: (3)*
Materials _____	
Supplies_____	_____
Equipment _____	_____
Services _____	_____
	_____

Timeframe: (4) _____
Personnel: (5)_____

(1) Annual increment of long-range objective, stated as the number of trees to be planted, pruned, removed, or otherwise treated, or the intended outcome of another activity.
(2) List and describe materials, supplies, additional equipment (may be rental), and services necessary to achieve goal.
(3) Costs must be based on current estimates or quotations.
(4) Indicate months when activity will take place.
(5) Indicate by title individuals responsible for conducting activity.
Attach supporting materials, such as species lists, locations of specific activities, and work specifications.

FIGURE 5.1. Operational planning worksheet.

must be based on solid planning information and the best judgment of qualified individuals. Reflecting the needs as determined by planning, budgets are thus "zero-based" rather than inflation-adjusted figures from the past year. Budget approval, through the city's financial office and ultimately by the mayor and council, depends largely on two factors: adequate justification, and public support for the urban forestry program. Even though budget approval is an annual event, it is the result of a total ongoing planning process plus continuing success in gaining citizen support.

PLANS OF WORK

Plans of work are internal departmental documents—tasks and timetables necessary to carrying out operational plans—listing what is to be done, when, and by whom. Plans of work do the following:

* Provide a view of the overall scope of work
* Allow personnel to see their roles clearly
* Help to ensure that nothing is overlooked
* Facilitate transition in case of personnel changes

There should be an overall departmental plan of work. Individual personnel should also have plans of work—often simply pocket or desk calendars with tasks entered.

FEEDBACK AND EVALUATION

Program evaluation is centered on annual goals, and goals are centered on long-range objectives. Were goals met on time and within budgets? If so, was the process as efficient and effective as possible? If goals were not met, why not? Evaluation should be both ongoing and reflective. Ongoing evaluation requires close monitoring, allowing deficiencies to be recognized and possibly corrected while programs are in progress. Reflective evaluation is generally at the end of the program year, when cost accounting is complete and all other information is available. Procedures for evaluation may be formalized by writing them into operational plans, or they may be separate as a part of departmental policy.

EMERGENCY PLANNING

No matter how well planned, urban forestry programs may have to change suddenly because of a natural disaster—a hurricane, tornado, ice storm, fire, flood, or even earthquake. In such events, long-range and operational plans may have to be reordered or replaced entirely. A hurricane or severe ice storm, for example, can render all baseline information about tree condition and needs invalid. This was perhaps best expressed by an urban forester who, when asked about the effects of a storm on the urban forestry program, replied wistfully, "In a matter of minutes we had gone from being caught up on our pruning to suddenly facing two years of work." (Fazio, 1988). Of immediate concern, though, will be the need to respond to new priorities thrust upon the urban forest because of the disaster. And the new priorities will be, in order, safety, clean up and salvage, and repair.

Public safety must first be ensured. Downed power lines, obstructing debris, and hazards caused by broken or weakened trees and branches must be taken care of first.

Clean up of fallen trees and branches is the next priority. At this time, it may also be possible to salvage occasional trees by rudimentary pruning, guying, or other practices.

Repair work may follow, perhaps over a period of months. In most cases, trees cannot be restored to their condition before the disaster, but they can be put in shape through corrective pruning and other arboricultural practices to develop functionally.

An emergency plan is needed for the urban forest. The plan must be in cooperation with other departments of local government and should consider the following:

- Communication capability with police, fire, health, and public utility officials
- Coordination of city, utility company, and private arborists
- Cooperation with local radio, television, and other media for announcements about safety, clean up, and tree salvage
- Disposal of debris
- Situation assessments
- Continuing education of property owners about tree repair and replacement planting

Many cities have overall disaster plans, where various scenarios are envisioned and sources of relief are identified. The urban forest should be an important consideration in such plans. An excellent reference to disaster planning is *Storms Over the Urban Forest: Planning, Responding and Regreening,* by J. W. Andresen and L. L. Burban (University of Illinois, Dept. of Forestry; USDA Forest Service; and Illinois Dept. of Agriculture, 1993).

SUMMARY

Planning for comprehensive management must begin with a long-range plan. The long-range plan states the mission, defines objectives, and identifies strategies and priorities for achieving objectives. Operational plans are based on the long-range plan and are incremental to achieving objectives. Budgets are based on operational plans. Plans of work are tasks and timetables for carrying out operational plans. Evaluation is the final and continuing aspect of the planning and budgeting process. Long-range planning is a function of the city forestry department and tree board, with input provided by key stakeholders in the urban forest. Operational planning, budgeting and work planning is a function of the city forestry department. Emergency planning is a separate process in cooperation with other departments of local government.

REFERENCES

Fazio, J. R. 1988. "When A Storm Strikes." *Tree City U.S.A. Bulletin No. 2,* p. 6. The National Arbor Day Foundation, Lincoln, NE.

Grey, G. W. 1994. *A Handbook for Tree Board Members.* The National Arbor Day Foundation, Lincoln, NE.

CHAPTER 6

Program Implementation

You gotta do it right.

Urban forestry program implementation begins with a knowledge of the technical aspects of arboriculture—how to correctly plant, prune, and do whatever else is necessary to maintain trees. Those physically responsible for such activities must possess this knowledge. From a managerial perspective, however, this knowledge needs to be applied to the forest rather than to trees. Responsible parties need to know how to organize and implement large-scale planting projects, how to provide for maintenance of the entire urban forest, and how to manage tree removals. This statement does not imply that urban forest managers do not need knowledge of basic arboriculture. It suggests only that knowing the standards of quality is more important than applying them directly. For example, it is more important for managers to know the difference between a correctly and incorrectly pruned forest than to be experts in pruning individual trees. Thus, the focus of this chapter is on the application of management to the total urban forest.

The urban forest has four fundamental needs: planting, maintenance, protection, and removal (Grey and Deneke, 1986). In this chapter each need is considered from the managerial perspective of a city forestry department with overall responsibility for the urban forest. Note, however, that, although the tasks necessary to meeting the needs are discussed separately, each is interrelated in terms of long-range objectives. For example, tree selection and pruning are both critical to the objectives of a safe and healthy urban forest. Thus, each task must be considered in context with ultimate objectives and relationships to other operations.

MANAGING FOR TREE ESTABLISHMENT

Even though the term *planting,* the first of the four fundamental needs of the urban forest, is essentially correct, the broader need is for tree establishment. Planting is more properly one of four steps in tree establishment: location and selection, planting, aftercare, and analysis. Hence, management of tree establishment within the urban forest must consider the various factors of location and selection of individual trees, the application of resources to planting, provisions for ensuring survival and health, and evaluation of the results.

Tree establishment is an ongoing need in the urban forest: to replace mortality, to supplement existing stands of trees, and to landscape new areas. At the outset, there are four overriding considerations. These four rules of urban forest tree establishment are:

1. Tree establishment must be done in accordance with the needs (numbers and locations) as identified in long-range plans.
2. Tree establishment must maintain or enhance diversity of the urban forest.
3. Trees selected must be consistent with the limiting factors of planting sites (soil, space, climate, other).
4. Trees selected must meet the remaining criteria as identified by objectives in long-range plans (safe, healthy, functional, pleasing, other).

Location and Selection

Few things are more important in urban forest management than matching trees with the sites where they are to grow and develop. Careful location and selection of trees is fundamental to meeting the long-range objectives for the urban forest. In fact, location and selection is a critical factor in meeting every objective—safe, healthy, diverse, well stocked, functional, and economical—for the forest. In addition, and extremely important, careful location and selection is a major factor in reducing the need for future maintenance, as is discussed later in this chapter.

Before elaborating, two points should be made. The first is that all "planting spots" do not need trees. There is much to be said for open space for its own sake and for its enhancement of visual and other values. The second point is that, all too often, tree selection occurs before location is considered. The rationale is: "I like a certain tree; therefore, I am going to put it there." Unfortunately, the urban landscape is replete with the results of such determination. With this said, let us continue.

With tree establishment needs quantified in long-range plans and incremental determinations made, general planting areas must be prioritized and trees then selected for specific sites. In most cities, budgets do not allow for all needed trees to be planted; thus, priorities must be determined. Ideally, trees ought to be planted where they will most economically provide the most benefits to the most people. Implementing such can be difficult because of community values and political considerations. Nevertheless, prioritization should begin with the question: Where should trees be located to provide the most benefits? The answer can be partially found in general land use categories—residential, commercial, and industrial—with further subcategories of each. For example, new residential areas without trees might take precedence over established neighborhoods where only replacement plantings are needed. Also, specific projects must be weighed against overall needs, as a disproportionate amount of total planting budgets might easily be consumed by establishing large trees in areas of high people use, such as outdoor shopping malls. Because of an environment of changing social, economic, and political circumstances, priorities must be flexible.

Three factors determine individual tree selection: the tree's purpose in the landscape, what the site will allow, and the amount of care required. Simply stated, the challenge of tree selection is to find the best low-maintenance tree for its purpose in the landscape that the site will allow. Harris (1983) summarized these factors as shown in Table 6.1.

Determining the function in the landscape is the first step in tree selection and location. The biggest question is: What is the primary purpose of the tree? Answers may include shade, wind protection, screening, enframement, accent, contrast, and wildlife attraction. The answer will not only influence the species to be considered but may also help determine the precise spot where the tree will be located.

Careful site analysis is a necessity in tree selection. Site analysis, sufficient for tree location and selection, can be done during inventories, or each site may be revisited. Most sites will impose constraints, narrowing the choices of trees that may be used. Fortunately, though, nature (with occasional help from human scientists) has designed trees with enough different characteristics to provide choices in nearly every site situation. These characteristics are considered following a discussion of site constraints.

TABLE 6.1. *Characteristics of Woody Plants and Suggested Relative Importance of Their Influence on Landscape Function, Site Adaptation, and Plant Care*

Plant Characteristics	Function			Adaptation to Site	Plant Care
	Architectural and Engineering	Climate and Human Comfort	Aesthetic		
Growth Habit					
Tree, shrub, vine	***	***	***	**	**
Size	***	***	**	**	***
Form	**	**	***	*	
Growth rate	*			*	**
Branching	**	*	*	*	**
Wood strength	**			*	*
Rooting	**			**	***
Plant Features					
Leaves	**	***	***	**	***
Flowers		*	***		*
Fruit			**		**
Bark			**		
Environmental Tolerances					
Temperature				***	**
Drought				**	**
Wind				**	**
Light				**	*
Soil				***	**
Air				**	*
Pests				***	***
Fire	*				**

*** = major influence; no * = little or no influence.

Site Constraints. Most sites have constraints to tree establishment, because of space, soils, and local environmental factors.

Spatial Constraints. Spatial constraints such as overhead wires, buildings, sidewalks, streets, driveways, signs, poles, and other trees are easily recognizable. Less obvious are underground utilities and other structures. In many cases, locational information about underground utilities is available from other departments of public works. Spatial constraints demand that trees be compatible with their environment during all stages of growth and development. Hence, it is necessary that growth, both size and form, be correctly anticipated. This vision is too often lacking in urban tree establishment. Trees must fit their allowable spaces and may be selected according to their various growth habits or their ease of pruning or otherwise molding to allow them to conform.

Soil Constraints. Urban soils are reputed to be "terrible," often perceived to be either severely compacted or polluted with physical or chemical materials. Also, nearly all trees are said to grow best in deep, fertile, well-drained soils. The latter statement, while in need of qualifying, is generally true. Most trees do thrive in good soils. Fortunately, urban soils do not always live up to their reputation. There are both favorable and unfavorable soils in nearly every city. Soil restrictions are generally caused by depth, compaction, pollution, low fertility, alkalinity, or acidity. Some species tolerate these conditions. The list of potential trees may also be broadened by correcting or amending soil deficiencies. Urban forest managers must, however, assess the cost effectiveness of corrective measures.

Local Environmental Constraints. The local environment sometimes imposes additional site constraints. These constraints may be microclimatic, involving shade, sun, and wind exposure, or they may be related to air or soil pollution.

Depending on degree, shade from buildings and other trees can be a limiting factor in tree selection, making shade tolerance an important consideration. Many species do not tolerate shade. Shade-tolerant species are those that, in their natural habitat, thrive as understory specimens beneath the higher forest canopy. Urban forest managers should consider the natural environment of each species during tree selection.

For practical purposes, sun and wind exposure may be considered as the opposite of shade tolerance. Shade-tolerant species often do poorly on sites exposed to hot sun and drying winds. A common example is the sugar maple. Selected primarily for fall color, sugar maples are often planted as core shade specimens on exposed building lots. In regions of high evapotranspiration, such as the Great Plains, such exposed trees have relatively little chance of becoming established. The influence of exposure is well illustrated by small cities in the western United States where sun- and wind-tolerant species such as the Siberian elm and black locust have become well established, contributing to a microclimate where less sun-tolerant species, including the sugar maple, have been successfully introduced.

Pollution, particularly from salt, may be a limiting factor in tree selection. Most applicable to northern cities where snow removal is a necessity and to seaside areas where ocean sprays are common, salt can injure trees and other vegetation. Injury can be from airborne particles or from excessive salt in the soil. Injury from airborne particles is mainly to leaves; thus, evergreen trees, especially those with little wax on their leaves, are more susceptible, both in northern areas where salt application is in the dormant season for deciduous trees and in seaside areas where exposure is year-round. Naked buds and tender stems can also be injured. Deciduous trees with resinous buds, such as the horse chestnut and cottonwood, and trees with submerged buds, such as the honeylocust, are resistant to injury (Lumis, Hofstra, and Hall, 1973). Injury from accumulated soil salt is often less severe, as the effects are insidious over time. The salt tolerances of various species follow (Pirone et al., 1988—adapted):

> *Very tolerant*—black cherry, redcedar, red oak, white oak
> *Tolerant*—black birch, black locust, cottonwood, gray birch, honeylocust, horse chestnut, largetooth aspen, paper birch, white ash, yellow birch
> *Moderately tolerant*—American elm, linden, hop hornbeam, Norway maple, red maple, shagbark hickory
> *Intolerant*—beech, hemlock, red pine, speckled alder, sugar maple, white pine

Tree Characteristics. As stated previously, nature, with occasional human help, has endowed certain trees with characteristics that make them suitable for the urban environment. Planting sites impose constraints. Nature provides opportunities to fit the constraints. The managerial challenge is to make the best possible match between trees and sites. Characteristics of trees allowing such matches to be made are climate adaptability; shade, exposure, soil, and pollution tolerance; insect and disease resistance; and size and form.

Climate Adaptability. Trees are in nearly every city, thereby testifying to their ability to withstand the rigors of local climates. For any given city, however, many tree species cannot survive there. The primary limiting factor is lack of cold hardiness. Cold hardiness zone maps are helpful in determining which species will grow in particular regions of the United States and Canada. The maps are limited, however, in that hardiness zones are based on average minimum temperatures and do not take into account the extremes that Mother Nature occasionally throws in. They are also deficient in that they do not consider drastic temperature changes, such as sudden cold following a warm period, that can cause severe injury to trees. Also, local topographical or other landscape features can cause microclimates quite different from zone averages. In spite of their shortcomings, hardiness maps can be very helpful, particularly when new, exotic species are considered. Figure 6.1 is a plant hardiness zone map. Lists of trees adapted to each zone are generally available from nurseries.

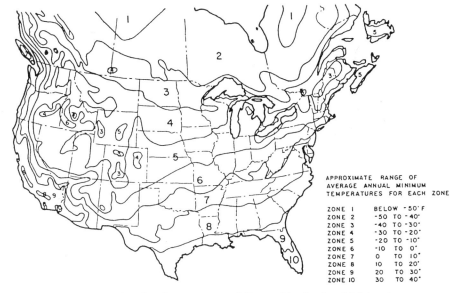

FIGURE 6.1. Plant hardiness zones as delineated by the Arnold Arboretum.

Shade Tolerance. Discussed previously as a site factor, shade tolerance is an important, and often overlooked, consideration in tree selection and location. Table 6.2 gives relative tolerance to shade of several
North American species (Aber, 1990).

Exposure Tolerance. Also discussed previously as a site factor, the
managerial need is to identify species that can become established if
planted on sites exposed to hot sun and drying winds. Please refer also
to Table 6.2.

Soil Tolerance. Of primary concern is identification of trees with
the ability to tolerate various soil conditions, such as compaction,
dryness, acidity, and alkalinity. Compacted soils are frequent in most
urban areas, with the attendant problems of drainage, aeration, re-
stricted gaseous exchange, and other conditions detrimental to tree
survival and growth. Such soils often require special treatments, such
as aerating or amending. The need for such treatments can be de-
creased, however, by selecting species with inherent tolerance to these
conditions. Observing nature gives obvious clues—species of willow,
alder, gum, and baldcypress growing in bogs and swamps and pin oak,
sweetgum, and pecan thriving in areas of poor drainage. Observing
trees after prolonged flooding is also helpful in identifying species
tolerant to poor soil drainage. Table 6.3 is a list of species that withstood

TABLE 6.2. *Relative Tolerance to Shade of Some North American Species*[a]

Very Tolerant

Eastern Conifers	Eastern Deciduous	Western Conifers	Western Deciduous
Balsam fir	American beech	Western redcedar	
Eastern hemlock	American hornbeam	Silver fir	
	Flowering dogwood	Western hemlock	
	American holly	California torreya	
	Eastern hop hornbeam	Pacific yew	
	Sugar maple		

Tolerant

Eastern Conifers	Eastern Deciduous	Western Conifers	Western Deciduous
Northern white cedar	Rock elm	Alaska yellow cedar	California laurel
Red spruce	Blackgum	Incense cedar	Canyon live oak
White spruce	Sourwood	Port Orford cedar	Tanoak
	Red maple	Grand fir	
	Hickory spp.	Subalpine fir	
		California red fir	
		White fir	
		Mountain hemlock	
		Redwood	
		Engelmann spruce	
		Sitka spruce	

Intermediate

Eastern Conifers	Eastern Deciduous	Western Conifers	Western Deciduous
Eastern white pine	Ash spp.	Douglas fir	Red alder
Black spruce	Basswood	Monterey pine	
	Sweet birch	Sugar pine	
	Yellow birch	Western white pine	
	Buckeye	Blue spruce	
	American elm	Giant sequoia	
	Sweetgum	Noble fir	

Baldcypress
Loblolly pine
Pitch pine
Pond pine
Red pine
Shortleaf pine
Slash pine
Virginia pine

Hackberry
Cucumber magnolia
Silver maple
Black oak
Northern red oak
Southern red oak
White oak

Paper birch
Butternut
Catalpa spp.
Black cherry
Chokeberry
Kentucky coffeetree
Honeylocust
Pin oak
Scarlet oak
Pecan
Persimmon
Yellow poplar
Sycamore

Intolerant

Bigcone Douglas fir
Juniper spp.
Bishop pine
Coulter pine
Jeffrey pine
Knobcone pine
Limber pine
Lodgepole pine
Piñon pine
Ponderosa pine

Madrone
Bigleaf maple
Oregon ash
California white oak
Oregon white oak
Golden chinquapin

Jack pine
Longleaf pine
Sand pine
Eastern redcedar
Tamarack

Aspen spp.
Gray birch
River birch
Post oak
Turkey oak
Blackjack oak
Willow spp.

Very Intolerant

Alpine larch
Western larch
Bristlecone pine
Digger pine
Foxtail pine
Whitebark pine

Quaking aspen
Cottonwood spp.
Willow spp.

Source: Adapted from H. W. Hocker, Jr., Introduction to Forest Biology, John Wiley & Sons, New York, 1979.
[a] Arranged in order of tolerance among groups but not within groups.

TABLE 6.3. *Flood-Tolerant Woody Species* [a]

Botanical Name	Common Name
Acer negundo	Boxelder
Acer rubrum	Red maple
Acer saccharinum	Silver maple
Carya aquatica	Water hickory
Carya illinoensis	Pecan
Carya ovata	Shagbark hickory
Cephalanthus occidentalis	Buttonbush
Cornus stolonifera	Redosier dogwood
Crataegus mollis	Red hawthorn
Diospyros virginiana	Persimmon
Eucalyptus camaldulensis	Red gum
Forestiera acuminata	Swamp privet
Fraxinus pennsylvanica	Green ash
Gleditsia aquatica	Water locust
Gleditsia triacanthos	Honeylocust
Ilex decidua	Deciduous holly
Liquidambar styraciflua	Liquidambar
Nyssa aquatica	Water tupelo
Planera aquatica	Water elm
Platanus × *acerifolia*	London planetree
Platanus occidentalis	Sycamore
Populus deltoides	Eastern cottonwood
Quercus bicolor	Swamp white oak
Quercus lyrata	Overcup oak
Quercus macrocarpa	Bur oak
Quercus nuttallii	Nuttall's oak
Quercus palustris	Pin oak
Salix spp.	Willow
Salix alba var. *tristis*	Golden weeping willow
Salix exigua	Narrow-leaf willow
Salix hookeriana	Hooker willow
Salix lasiandra	Pacific willow
Salix nigra	Black willow
Taxodium distichum	Baldcypress
Ulmus americana	American elm
Washingtonia robusta	Mexican fan palm

[a]The species listed withstood 180 or more days of water covering the soil under trees. Observations were made in the ten U.S. Army Corps of Engineers Divisions in the contiguous United States.

180 or more days of water covering the soil under the trees (Whitlow and Harris, 1979).

Pollution Tolerance. As indicated in the section concerning salt pollution, there is a strong need for trees with tolerance to both physical and chemical pollutants. Physical pollutants, primarily dust, impair plant functions by restricting respiration or other growth processes. Chemical pollutants may cause physical injury, but they also influence the chemical processes within plants. In some locations within the urban forest, pollution from salt, sulphur dioxide, ozone, or other source is a certainty, and trees with resistance must be selected accordingly. In many other areas, pollution will be random, resulting mainly from accidents or misuse of chemicals. Tables 6.4 and 6.5 give relative sensitivities of selected species to sulfur dioxide and ozone (Loomis and Padgett, 1975).

Insect and Disease Resistance. To select insect- and disease-resistant trees requires a knowledge of current local plant problems. Selection must also anticipate future problems, such as the spread of gypsy

TABLE 6.4. *Relative Sensitivity of Selected Forest Species to SO$_2$*

Sensitive	Tolerant
American elm	Black tupelo
Ash	Boxelder
Aspen	Dogwood
Birch	Juniper
Blackberry	Live Oak
Blackjack oak	Maple
Carelessweed	Sourwood
Catalpa	Spruce
Dewberry	Sycamore
Eastern white pine	Yelow poplar[a]
Jack pine	
Larch	
Loblolly pine, (seedlings to 6 ft)	
Poplar	
Ragweed	
Virginia pine, (seedlings to 6 ft)	

[a]Sensitive spring and early summer.

TABLE 6.5. *Relative Sensitivity of Selected Forest Species to Ozone*

Sensitive	Tolerant
Ash	Balsam fir
Eastern white pine	Black walnut
European larch	Gray dogwood
Honeylocust	European white birch
Jack pine	Maple
Poplar	Red oak
Sweetgum	Spruce
Sycamore	
Virginia pine	
White oak	
Yellow poplar	

moth infestations. Assistance is available from plant health specialists from the Cooperative Extension Service and state forestry agencies.

Size and Form. Spatial constraints require that trees conform. Conformity of established trees can be attempted mechanically by crown and root pruning or chemically by applying growth inhibitors. Both methods are expensive, and neither has lasting results. Fortunately, nature gives urban forest managers some opportunities at planting time to prevent or reduce the necessity of these measures. Trees come in different sizes and shapes.

Trees generally can be grouped in three mature size classes: Small, Medium, and Large. Maximum mature heights are 30 feet for Small Trees, 60 feet for Medium Trees, and 100 feet for Large Trees. The three classes serve as guides in selecting trees that may conform naturally to spatial constraints. Even though size classes are extremely useful tools, you must understand that they may be inexact. They are based on average heights at maturity and also do not take into account the influences of individual sites on ultimate growth. A selected list of species by size classes follows.

Small trees	*Medium trees*	*Large trees*
Apricot	English oak	Bur oak
Bradford pear	Green ash	Cottonwood
Flowering crabapple	Hackberry	Kentucky coffeetree

Flowering peach	Honeylocust	London planetree
Goldenrain tree	Japanese pagodatree	Silver maple
Hawthorn	Linden	Sugar maple
Japanese tree lilac	Osage orange	Sycamore
Purpleleaf plum	Pecan	
Redbud	Persimmon	
Serviceberry	Red mulberry	
	Red oak	
	River birch	
	Sassafras	
	Soapberry	

Under normal growing conditions, all landscape trees have a characteristic form, or shape, created by branch structure and growth habit. Tree forms not only provide opportunities for fitting trees in physical spaces but also constitute an important landscape design element. Although forms sometimes change as trees mature, tree form is such a strong characteristic of many species that it is an important factor in tree identification. Trees assume seven general forms: irregular, vase, oval, pyramid, fastigiate, round, and weeping (Figure 6.2). These forms are sometimes refined by applying the words *broadly, moderately,* or *narrowly* to each. For example, a tree may be broadly oval, moderately oval, or narrowly oval.

The importance of tree form is perhaps best illustrated by streetside trees. Generally, the most suitable form is vase, where branches ascend normally to provide lateral clearance. Irregular, oval, round, and fastigiate forms may also be used. Although often used, pyramid and weeping forms are seldom suitable for streetsides, as they create visual and physical obstructions for movement of vehicles and pedestrians. To remove such obstructions, heavy pruning of lower branches is needed, often resulting in reduced health and vigor. Common examples of pyramidal streetside trees are pin oak, willow oak, and sweetgum. Pyramidal trees do have their place, however, in certain street environments. They serve admirably, for example, in wide boulevards where their lower branches provide screening from oncoming headlights.

IRREGULAR   VASE   OVAL   PYRAMID   FASTIGIATE   ROUND   WEEPING

FIGURE 6.2. Characteristic shapes of landscape trees.

Negative Characteristics. Although few trees are without problems throughout their lifespans, some trees have characteristics that cause public dissatisfaction. Miller (1988) refers to such negative qualities as social externalities.

Excessive fruiting, prolonged heavy leaf fall, shredding bark, root suckering, and bad odors all make a species undesirable. Species that attract nuisance wildlife and other pests, such as aphids, which drip honeydew on automobiles, should be avoided, as should species with dense, leafy crowns that shade out and kill lawns.

Economics. Nearly all the preceding factors have economic elements, relating to the costs of establishment and maintenance and suggesting that such considerations are a primary driving force in tree selection. Economic analysis is a necessity. Initial cost of planting stock and subsequent establishment costs of various species must be compared with establishment rates. Such evaluation will reveal whether higher-cost planting stock—because of species, size, or type (balled and burlapped, containerized, or bareroot)—is more or less economical. To make such comparisons, an establishment period must be chosen (generally 3 to 5 years), all costs recorded, and establishment rates (percentages of live, healthy trees) determined. Table 6.6 provides an example, comparing the establishment costs of balled and burlapped redbuds with bareroot redbuds. Tree establishment analysis is discussed further at the end of this section.

TABLE 6.6. *Comparison of Establishment Costs of Balled and Burlapped and Bareroot Redbud Trees* [a]

	Balled and Burlapped	Bareroot
Initial purchase price per tree ($)	32.00	14.00
Planting cost per tree ($)	18.00	15.00
Maintenance costs (3 years) ($)	8.00	8.00
Total establishment cost per tree ($)	58.00	37.00
Number of trees planted	150	150
Total establishment cost ($)	8,700.00	5,550.00
Establishment rate (%)	75	50
Number of established trees	112	75
Cost per established tree ($)	77.68	74.00

[a]Although there is an obvious cost saving of $3.68 per tree by selecting bareroot trees, 37 additional trees will have to be replanted.

Other Considerations. Most of the preceding factors concerning tree selection and location are physical, dealing mainly with site constraints and tree characteristics. There are also issues of public health and safety to be considered, as well as attitudes and wishes of urban residents. Regarding health, trees notorious for pollen that causes allergies should be avoided. For safety reasons, trees prone to storm breakage and those with thorns, toxic fruits, and surface roots should be planted only where they do not present hazards. In the final analysis, trees must serve the public, healthfully, safely, functionally, and aesthetically. The factors involved in species selection are summarized by the model shown in Figure 6.3 (Miller, 1988).

A Final Note: Tree selection and location have recently gone modern. Computer programs considering all the variables discussed previously are now available, ultimately pictorially displaying recommended species. A model program is "Southern Trees," produced by the University of Florida, with funding from the USDA Forest Service.

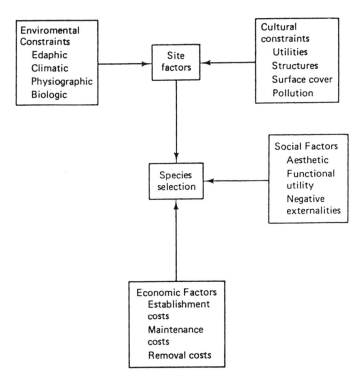

FIGURE 6.3. Species selection model. From Miller, Robert W. *Urban Forestry: Planning and Managing Urban Greenspaces,* © 1988, p. 188. Reprinted by permission of Prentice-Hall, Inc., Englewood Cliffs, N.J.

Planting

From a managerial standpoint, planting should be considered as a process, including everything necessary after species selection and site location to ensure that trees are securely in the ground. The planting process has four major considerations: quality assurance, contracting, site designation, and implementation.

Quality Assurance. When tree characteristics are carefully considered, the best species for each site can be chosen. What is next necessary is that the best-quality trees of each species be assured. High-quality and adaptable planting stock is critical to establishment, growth, and function in the landscape. Quality assurance is also a major economic factor—reducing replacement costs and future maintenance needs.

Planting stock sources are primarily commercial nurseries, although some cities grow their own materials, and in some cases trees are transplanted from native forests or tree plantations. Whatever the source, quality planting stock must be assured.

Quality standards for planting stock, as adopted by the American Association of Nurserymen (AAN), are detailed in *American Standards for Nursery Stock,* which is available from AAN (1250 I Street NW, Suite 500, Washington, DC 20005). All urban forestry managers responsible for tree establishment should be familiar with this publication, and its standards should be the basis for all planting stock acquisition.

Quality assurance begins with the ability to recognize good planting stock, in the nursery and at time of delivery. If possible, it is an excellent idea to go the nursery fields and look at the plant materials before seeking bids or before digging begins. In the field, questions such as the following relative to branches and trunk should be considered. During delivery inspections, the entire list should be considered (Grey, 1993).

* *Branches and trunk*

    Is the crown symmetrical, without evidence of branch ends being pruned?

    Have the lower branches been excessively pruned? (Two thirds of the total height should be in live crown.)

    Are branches well distributed around the trunk and considerably smaller than the trunk?

    Is the tree free of broken branches?

    Is the tree free of crossing and rubbing branches?

    Is the trunk well tapered?

    Is the tree free of insect, disease, or other damage?

    Did the tree grow vigorously during the preceding season?

- *Bareroot tree*
  Is the mass of the roots proportionate to the trunk and crown?
  Are the roots symmetrical?
  Are the roots moist and fibrous?
- *Balled and burlapped tree*
  Is the root ball adequate to the tree's size?
  Is the root ball solid and intact?
  Is the root ball moist?
  Is the material around the ball really burlap or is it, and the string, a synthetic fiber? (If synthetic fiber, it should be removed when the tree is planted.)
- *Container grown tree*
  Is the container adequate for the tree's size?
  Are the roots moist?
  Are there circling or girdling roots? (To determine this, a destructive sample may be required.)
  Is the tree really container grown or was it recently transplanted to a container? (Again, a destructive sample may be required.)

The source of seeds or propagational materials from which nursery stock is grown is also important. To ensure hardiness, planting stock should be from seed sources from the same climate zone as where it will be planted. This is of particular concern where stock grown from seed from a species near the southern edge of its native range is offered for planting in northern zones. Seed source considerations are in addition to the general climate adaptability factors discussed previously.

A final factor in quality assurance is how the planting stock is handled after delivery. A holding area where trees can be stored prior to planting is necessary. Such an area must be secure from theft, protected from hot sun and drying winds, and have a watering system. Mulch material must also be available to prevent root drying. Trees can be transferred daily to planting sites, with care being exercised in physical handling to avoid root ball breakage and other damage. Root ball breakage is a common occurrence, often causing traumatic damage resulting in failure of trees to survive.

Contracting. In many cities adjacent property owners are encouraged to plant streetside trees, and in some cities it is the only means of new tree establishment. In the interests of planned urban forestry programs, such plantings must be guided—at the minimum by official species lists (either recommended or prohibited species) and at the maximum by permits coupled with on site visits by city forestry department representatives. Permits should be specific as to future management responsibilities for trees planted.

On city property, trees may be planted by city crews, by volunteers, or by contract with commercial landscape firms. Decisions as to which approach is taken are based on factors such as the following:

Other work priorities for city crews
The cost of additional labor for planting
The handling of materials and supplies
The equipment cost and availability
The length of the planting season
Availability of volunteers
Past experience
Liability considerations

Volunteer involvement in planting has the obvious advantage of reducing labor costs. It also helps build program support, as volunteers often develop a sense of proprietorship in the trees they have planted and in the total urban forest. Working with volunteers is discussed in Chapter 7.

Contract planting has the advantage of relieving city forestry departments of the responsibility of planting details and reducing liability obligations. Contract planting may initially be more expensive, but it can be economical in the long-run because of survival guarantees.

Whether planting is done by city crews or volunteers, trees must be acquired from city nurseries or commercial nursery suppliers. Acquisition of tree planting stock from commercial nursery suppliers involves bid invitations and contracts. Procedures for inviting bids and entering into contracts are established by administrative policy within most cities. For planting stock, adequate lead time (usually several months) should be allowed, and requirements must be precise. Figure 6.4 lists the primary considerations in bid invitations and contracts for purchase of tree planting stock. Planting by landscape firms also necessitates bid invitations and contracts. Many of the elements are the same as for acquisition of planting stock, but other considerations such as identity of planting spots, guarantee conditions, and contract period must be included. Figure 6.5 lists considerations in inviting bids and entering into contracts for tree planting.

Site Designation.   Precise spots where each tree will be planted must be designated, with assurance that the correct tree will be planted in each spot. Site designation begins with a review of underground utility information considered during tree selection and location. Urban forest managers should be aware that in some states excavators, including tree planters, are required by law to give notification of planned excavations. An example is Virginia where a minimum of 48-hour notification is required. Public utilities must

1. Names, addresses, and telephone and fax numbers of involved parties.
2. Species and varieties (common and scientific names).
3. Number of trees by species. (Nurseries may give quantity discounts.)
4. Size of planting stock by height intervals, stem caliper (diameter 6 inches above root collar), or container size.*
5. Condition of planting stock: Specified as bareroot, balled and burlapped, or container grown. Statement that all stock must be healthy, vigorous, and well grown.*
6. Source: Requirement that all planting stock be grown from seed or other propagational materials from the same climate region as where it is to be planted.
7. Available date: Weather and other conditions may influence delivery. Hence, a time period such as "first week in March" should be stated and then confirmed as time draws near.
8. Prices: Quoted by species, size, and condition; delivered to a specific location or FOB nursery; and time and method of payment.
9. Guarantees: Nurseries generally guarantee their stock to be in good growing condition and will replace only if mortality is their fault.
10. Substitutions: Allowances for substitutions of species, sizes, and conditions if planting stock as specified is not available.
11. Inspection and right of rejection:
    Nursery Inspection—Statement that city forester has the right to inspect and tag stock before digging.
    Agency Inspection—Requirement that all stock be certified as insect and disease free.
    Delivery Inspection—Statement that city forester may inspect all stock upon delivery.
12. Receipt of stock: Whether stock is to be picked up or delivered. If delivered, include the precise location.
13. Dates of bid submission and/or contract entry.
14. Signatures.

Please Note: For size and condition, it should be stated that all planting stock must conform with the American Standards for Nursery Stock, as adopted by the American Association of Nurserymen.

FIGURE 6.4. Considerations in bid invitations and contracts for purchase of tree planting stock.

determine whether lines are in the way and, if so, mark them clearly. With utilities located, planting spots can then be marked. Marking can be by wooden stakes driven into the ground, with a species code lettered on each stake. A commonly preferred method, however, is to mark the species code directly on the ground with spray paint. Marking should be done slightly ahead of planting, as stakes may be pilfered or otherwise moved, and paint may fade. Planting crews must be familiar with species codes.

1. Names, addresses, and telephone and fax numbers of involved parties.
2. Number of trees to be planted.
3. Species and varieties to be planted (common and scientific names).
4. Size of planting stock by height, stem caliper, or container size.*
5. Condition of planting stock: Bareroot, balled and burlapped, or container grown.*
6. Source: Requirement that all planting stock be grown from seed or other propagational materials from the same climate region as where it is to be planted.
7. Substitutions: Allowances for substitutions of species, sizes, and conditions if planting stock as specified is not available.
8. Planting locations: By streets, blocks, parks, or other legally identifiable areas.
9. Planting requirements: Site preparation, backfill material, staking, mulching, pruning, other.
10. Work requirements: Statement that all planting activities will be performed during normal working hours and that no materials, supplies, equipment, or open planting holes will be left at sites after working hours.
11. Planting period: Beginning and ending dates when planting will be accomplished.
12. Replacement guarantees: Requirement (agreement) that all dead or severely stressed trees (excluding causes beyond the control of the contractor) will be replaced within a certain time period.
13. Price: Quoted by species, size and condition in bid submission. Quoted by total price in contract.
14. Inspections: Right of city to inspect planting stock prior to planting and to inspect trees during contract and guarantee periods.
15. Indemnification: Statement that contractor will hold city liableless in event of injury to life or property.
16. Performance bonding requirements.
17. Special exemptions: Inclement weather, natural disasters, civil acts endangering contractor's representatives, other.
18. Conditions of contract termination.
19. Dates of bid submission or contract entry.
20. Signatures.

Please Note: For size and condition, it should be stated that all planting stock must conform with the American Standards for Nursery Stock, as adopted by the American Association of Nurserymen.

FIGURE 6.5. Considerations in bid invitations and contracts for tree planting.

Implementation.   Implementation of tree planting has both physical and public relations aspects. The physical aspect involves logistics of plant materials, equipment, supplies, and labor, with the challenge of efficiency, effectiveness, and economy. Quality control is imperative, ensuring that continuing attention is given to plant materials handling, hole digging, back filling, mulching, and other planting details. The public relations aspect is both internal and external. Internal public relations involves coordination with other city departments concerning other activities where planting is to be done such as street repair, utility work, and other restrictions to access. External public relations are extremely important to help to ensure survival and establishment of newly planted trees and to gain public support for the overall urban forestry program. In cases of streetside plantings, adjacent property owners or residents should be personally contacted, if practical, telling them that new trees are to be planted and asking them to look after the trees. A flyer with information about the species and its care can be presented. Doorhangers giving such information and thanking residents in advance for their attention to the new trees should be used when personal contacts cannot be made.

Tree planting often offers excellent opportunities for news stories. A media announcement of the city's overall plans for tree planting, stressing the benefits for the future, can be a positive item amid the often otherwise gloomy news of the day. Also, specific planting events, particularly those involving volunteers, can be of media interest. In such cases, focus should be on the volunteers, helping in an overall plan to make the city a better place to live. Working with the media is discussed in Chapter 7.

Aftercare

The preceding discussion has been directed toward getting the best trees secured in the ground, with good assurance of establishment. To further ensure establishment, aftercare is needed to provide for health and vigor—such things as water management, insect and disease management, and prevention and correction of mechanical problems. Many tasks are necessary: watering, mulching, removing staking and guying materials,* monitoring insects and diseases, pruning, and repairing wounds. And someone has to do each task.

An established tree may be defined as one that under normal conditions may thrive on its own after a given period of time. Establishment periods vary by species and growing sites, but on the average are 2 to 4 years. During this period, aftercare must be provided. An important step toward aftercare, as mentioned previously in the case of streetside trees, is getting adjacent residents to assume partial

---

*The trend is toward less staking and guying. Staking is costly, can be detrimental to trees, and in most cases is unnecessary. Some trees must be staked, however—those on extremely windy sites and those in areas of heavy public use where staking materials provide barriers to physical damage.

responsibility. Such is not only a cost savings to city forestry departments but also helps create a sense of ownership in the urban forest. The difficulty, however, is that care may be sporadic and misapplied, as some residents will do nothing, and many will not know how to correctly treat trees. Thus, resident involvement is often sought only for watering or other basic treatments. Watering by residents is particularly important, since watering by city crews or contract applicators (especially if water must be trucked to the site) can be excessively expensive.

Some aftercare treatments, such as mulch reapplication and stake removal, are predictable and can be scheduled. Other necessary measures—watering because of drought conditions, insect and disease treatment, and mechanical damage repair—may not be predictable. Hence, monitoring is necessary. Monitoring is most economical if accompanied by treatment capability. For example, if a tree is found to have broken limbs because of vandalism, the person monitoring should do corrective pruning, eliminating the necessity for a second visit to the tree. Aftercare treatments should be recorded, as such records can be an important part of tree establishment analysis.

Analysis

Tree establishment analysis can be extremely important in making future tree-planting decisions. Analysis allows economic comparisons to be made among species, by sizes of planting stock, by condition of stock (bareroot, balled and burlapped, and container grown), among nursery sources, among planting sites, and by management activities. For example, with analysis the urban forestry manager can know whether less expensive bareroot stock of a particular species is more or less economical in the long run than more costly balled and burlapped stock; whether tree staking is economical; or whether less costly smaller stock will do just as well as larger material. Analysis will also answer whether in-service or contract establishment is more economical.

Analysis need not be complicated, but accurate records must be kept, as it is necessary to calculate total establishment costs per tree. Total costs should include the original purchase of trees and everything done to them to ensure establishment—planting, mulching, pruning, staking, stake removal, watering, protecting, and such. Also, the reasons for each treatment should be recorded, as patterns may become apparent (for example, vandalism in certain areas) that may influence future planting.

MANAGING MAINTENANCE

Although an imperfect term, *maintenance* includes everything necessary between establishment and removal to ensure the optimum function of the urban forest. There are three basic, but interrelated, areas of maintenance management in the urban forest: hazard management, plant health management, and quality improvement. As will become apparent, a major part of maintenance involves pruning, because it is primarily what must be done to reduce hazards and improve the quality of trees. Pruning is also a strong factor in plant health management. Pruning management will be treated under quality improvement.

Hazard Management

Hazard management is discussed first because reduction of threats to life and property must always be top priority in every urban forestry program. I make this statement even though I know that some cities have no tree hazard management programs. In some cities, tree hazards are so prevalent that the cost of correction might well exceed the cost of potential liability claims. Tragically, in these cities, because of severe budget deficits and other problems, there are no plans for tree hazard management. It appears inevitable that damages will occur and that both civil and criminal legal actions will follow.

Liability is based on tort law, requiring prudent and reasonable exercise of care. In the urban forest, owners of trees (managers in the case of trees on public property) are liable for damages if reasonable care was not exercised. The consequences of tree hazards can be tragic, involving property damage, bodily injury, and even human death. Pain and suffering can be severe, and legal compensation can be staggering. The basic legal question in cases of liability litigation is whether reasonable and prudent action was taken by urban forest managers prior to the casualty. Thus, tree hazard management must be of highest priority.

Tree hazards are most commonly thought of as limbs that might fall or entire trees that might topple during storms. The potential is much broader, however, and from a personal perspective includes the following:

- *What might fall on you*
  Branches
  Fruiting bodies
  Entire trees

- *What might cause something to fall on you*
  Electric lines
  Other structures
- *What you might run into*
  Trunks
  Branches
- *What you might trip over*
  Roots
  Fallen branches
  Sidewalks heaved by roots
- *What you might slip on*
  Leaves
  Fruit
  Flower petals
- *What might obstruct your view*
  Trunks
  Branches

As suggested by the personalized list, two elements are necessary for tree hazards: trees predisposed to any of the preceding and the presence of someone or something of value. For a falling tree or falling parts of a tree to be a hazard, there must be a "target" (people, vehicles, structures, or other objects of value) within striking distance. (See Figure 6.6.) In all other cases, there must be direct contact by people (either as pedestrians or in vehicles).

Although discussed here as a separate topic, tree hazard management should not be considered as a separate program but should be a part of overall operations, interwoven with all management activities. Tree hazard management has the sole objective of preventing casualties. There are two alternatives: make trees safe or remove targets. The latter is sometimes necessary, particularly in cases of large, decadent historic or specimen trees where pruning, bracing, and other arboricultural measures cannot ensure safety. Target removal is accomplished by fencing, signing, or otherwise prohibiting access to a tree's strike zone. Less direct methods such as redirecting pedestrian traffic and moving portable picnic tables from the vicinity of suspect trees may also be used. Such measures are applicable in special cases, however, and cannot apply to the general urban forest.

It is the former (making trees safe) that requires a continuous commitment in urban forest management. There are two aspects to making trees safe: prevention and correction. Prevention involves selecting and establishing new trees with less potential to be unsafe and applying arboricultural practices to ensure health, vigor, and strength.

FIGURE 6.6. For a hazard to exist, there must be potential for tree failure and a target.

Correction involves identification and treatment of currently unsafe trees or residue from trees. Both, however, require a commitment to an action program. Figure 6.7 presents ten elements of a successful hazard management program. Correction will be discussed first, as priority must be given to existing unsafe conditions.

Correction. Correction first requires recognition and anticipation of potential hazards. Recognition applies to current unsafe situations, such as branches predisposed to failure, trees likely to fall, heaved sidewalks, and view-obstructing branches. Anticipation refers to predictable (generally seasonal) events such as fallen leaves, fruit, and other materials that may be hazardous to pedestrians or motorists. Recognition

1. Develop a public education program, in conjunction with an internal hazard tree training program, to involve the public in detection of hazard trees and to obtain public support for treatment or removal of defective trees in the interest of public welfare.
2. Establish a tree hazard policy with clearly defined objectives.
3. Assign an inspector who has responsibility and authority to act on established policies.
4. Conduct staff training within your department and coordinate with other departments that have contact with trees to watch for and report hazard trees.
5. Provide professional training in identifying and evaluating trees for potential hazard.
6. Establish a systematic inspection. Stratify the tree population into different age, size, and species classes and inspect first those trees most likely to create a hazard. Be systematic in the inspection of each tree.
7. Develop a checklist for visual inspection.
8. Provide opportunities on the job to validate a diagnosis. Encourage dissection of defective trees and their parts to continually learn and to improve skills.
9. Provide adequate documentation by consulting with city legal and administrative departments. These documents may be subpoenaed if legal problems arise.
10. Take action based on the inspection. Apply common sense when prioritizing work needed. Act quickly in high-risk situations. Close or barricade areas if necessary.

Source: USDA Forest Service.

FIGURE 6.7. Ten counts to a successful tree risk management program.

requires inspection, and inspectors must have knowledge of tree physiology and be able to recognize indicators and signs of potential tree failure. Some indicators of failure are obvious—dead or hanging branches, heaved roots, lightning or mechanical injury, and such. Less obvious indicators are internal trunk and branch decay and various root problems. Listed below are some things to look for. It is recommended that this list be supplemented by information from the following sources: *Tree Hazards—13 Questions That Could Save Your Life* (leaflet), *A New Tree Biology* (book), *Tree Biology and Tree Care* (book) (available from Shigo and Trees, Associates, 4 Denbow Rd., Durham, NH 03824), and *Tree Health Management: Evaluating Trees for Hazard* (videotape) (produced by the USDA Forest Service and available from the International Society of Arboriculture, P.O. Box 908, Urbana, IL 61801).

Species. Some species such as willows, silver maple, red maple, and boxelder are more brittle and prone to storm breakage. Additionally, such breakage often results in internal decay, further increasing the potential for failure.

History. What has happened to the tree in the past? Has it, for example, been topped, affording entry points for decay? Have its roots been severed by utility trenching or has the soil been excessively compacted by construction or other activity? Has the tree been recently pruned or otherwise treated? Has there been other activity? Check first.

General vigor. Check whether a tree is growing well and is in good health. Vigor is often reflected by leaf color and size and previous annual stem growth. Low vigor is not a hazard in itself but rather a symptom of a potentially unsafe tree.

Crown dieback. Smaller dead branches in a tree's crown can be hazards in themselves, but they are often indicators of other problems, particularly in the roots, possibly sufficient to cause the tree to fall.

Dead or dying trees. Whether trees are dead or dying is generally obvious. Dead trees are hazards only when there are targets. Leaving dead trees where no hazards exist can be beneficial to wildlife and other organisms.

Dead and hanging branches. Dead branches inevitably will fall. Hanging branches are those that have fallen on other branches or those that have broken but are not completely free from the "parent" stem.

Crossing and rubbing branches. Branches that cross and rub against one another may cause physical weakness or provide entry for decay.

Wounds. Wounds can both weaken a tree and invite decay. Of particular concern are wounds caused by lightning—splintered limbs and cracks in trunks, even extending into the ground. In most cases, internal damage is difficult to discern. If not removed, such damaged trees should be monitored closely.

Forked trunks. Depending on species, tightly forked trunks can be prone to splitting. Indicators of weakness are impacted bark, slime flux, and signs of decay.

Decay signs. Clues to internal decay are cavities, cankers, and fruiting bodies of fungi. Internal decay does not necessarily mean that a tree is unsafe, but it is a strong signal that there should be frequent inspections.

Balance. Although leaning and lopsided trees are generally more prone to fall than vertical trees, not all such trees are unsafe. The reasons for leaning trees should be determined. If a leaning tree has always grown that way, there is probably no problem. If, however, the lean is caused by a loss of support roots, there is almost certainly a hazard.

Inspections for hazardous trees may be done during inventories, during the course of normal work, after storms, and on a planned schedule. Total dependance should not be on observations made during inventories and normal work. Such observations should supplement scheduled inspections and those necessary following storms. Planned inspections should begin with historical reviews—from records, if available, of planting dates, all subsequent maintenance, any disturbing activities (such as trenching or other construction), and previous inspections. Ideally, all such information should be computerized, allowing easy access by location, species, age, and other factors. Inspection records and recommendations for action must be kept. Supported by tree condition factors, such recommendations may include the following:

- No action necessary
- Continued monitoring (suspect tree)
- Complete removal of tree, including stump
- Removal of specific branches
- Bracing and cabling
- Pruning to balance crown

Inherent in each recommendation is an indication of urgency. This should be made more specific, however, by statements such as "immediate action required" or "include in scheduled pruning." From a legal standpoint, if an "unreasonable" time elapses between identification of a hazard and a casualty, the urban forest manager can be held at fault. Thus, "reasonably immediate" action is called for. Note that ignorance of hazard trees does not absolve urban forest managers of responsibility—and liability.

Prevention. Involved in prevention are all management practices that may keep trees from developing hazards. As discussed previously, hazard prevention must be integrated with all other management functions and directed toward the objective of keeping the urban forest safe. Specific practices include:

Species selection. Hazard prevention begins with species selection, avoiding species known to be brittle or have other inherent problems that may make them unsafe. Important also is insistence on high-quality planting stock to ensure proper future branching structure and tree vigor.

Planting location. Directly linked to species selection, judicious location of trees can avoid making trees, by their presence, hazards and can reduce the necessity of costly measures to keep them safe.

Health care. Dedicated to keeping trees healthy, such measures as insect and disease management, maintenance of soil fertility, and water management can help trees overcome environmental factors that may make them unsafe.

Pruning. Pruning is extremely important in hazard prevention to develop strong scaffold branches, to eliminate crossing and rubbing branches, to discourage decay entry through broken branches, and to provide visual clearance for pedestrians and vehicles. Systematic pruning is a necessity.

Protection during construction. Prevention or minimization of damage to trees during construction, by measures such as rootzone protection and avoidance of trunk and branch wounding, can be of great value in reducing future hazards. Protection of trees during construction is discussed further in Chapter 7.

The importance of record keeping should again be emphasized. Ideally, each tree should have a complete recorded history, including selection, location, establishment, and every other treatment (including inspections) during its lifetime. In the event of a casualty, such information can be invaluable in determining that reasonable and prudent care was exercised.

Plant Health Management

Plant health management (PHM) has replaced integrated pest management (IPM) as the new standard in the arboricultural profession (Ball, 1994). As a concept, plant health management has its basis in ecology— the interrelationship of all factors of the biological environment of the urban forest. As a practice, plant health management requires an understanding of these relationships, with particular emphasis on soils, sites, physiological needs of plants, and life cycles of insect and disease pathogens. From a management standpoint, this means that health practices applied must consider all factors influencing a particular plant.

Plant health management also involves a tolerance factor that includes both plants and people. Plants may be able to tolerate "problems," particularly those caused by insects or diseases, with little but aesthetic damage, but people often will not tolerate the symptoms or the organisms that cause them. This does not mean that there are not situations in need of treatment, but it does suggest that treatments are often unnecessary for plant health. It suggests also a part of urban forest maintenance that is often excessively expensive. Plant health management should begin with the understanding that insects and other organisms that chew leaves, suck juices, and bore into stems are part of the natural environment, as are fungi, bacteria, and viruses. The urban forest manager has to deal with these situations in the context of their overall impact on plants, plus public tolerance of the symptoms, the pathogens, and the necessary control measures, especially if chemicals are involved.

The objective of plant health management is maintenance of health and vigor so that impacts of pathogens will be minimized and the necessity of direct treatments, particularly by chemical pesticides, will be reduced. Plant health management includes the following:

Selection and location
Improvement of the growing environment, especially soil improvement (fertility, drainage, aeration)
Pruning and wound control
Environmental controls of insect populations or disease infestations
Pesticide alternative controls
Pesticide application
Public education
Monitoring and recording

In 1993 the USDA Forest Service established a National Center of Forest Health Management in Morgantown, West Virginia. Although not specific to urban forestry, much of the Center's work has application to the urban environment: biorational methods using natural agents such as pheromones or microbials to neutralize damage from forest insects and pathogens, biological controls using parasites and predators of forest pests to prevent or mitigate pest activity, and nontarget effects of pest management actions (Society of American Foresters, 1994). Urban forest managers may get on the Center's mailing list by contacting the USDA Forest Service, National Center of Forest Health Management, 180 Canfield Street, Morgantown, WV 26505.

Quality Improvement

In a larger sense, comprehensive management is all about the improvement of the quality of the urban forest (and its ultimate function) and all the management practices previously discussed to enhance benefits are designed to meet this end. Even though this is true, quality improvement can be considered as a separate topic, particularly as it applies to individual trees. Improvement of the quality of individual trees is accomplished largely through pruning. Few landscape trees in the urban environment escape the need for pruning—for hazard prevention, for structural development, for health and vigor, and for physical and visual clearance. Pruning is, in fact, the largest single item in most city forestry program budgets. Pruning is an ongoing need, as natural growth processes and various environmental factors make it necessary, or advisable, perhaps several times during a tree's lifespan.

Yamamoto (1985) has identified five management approaches to the pruning of municipal trees: request pruning, crisis pruning, task pruning, species pruning, and programmed maintenance. As suggested by the title, request pruning is in response to adjacent property owners or residents asking that trees be pruned. Crisis pruning involves removing immediate hazards. Crisis pruning, as applied to individual trees, must always be done as hazards are identified. For such to be the basis of a city's total pruning, however, is unfortunate evidence of a far less than adequate urban forestry program. Task pruning is for specific reasons other than immediate hazard removal such as removing lower branches overhanging sidewalks to allow for pedestrian passage. Species pruning allows for particular needs of certain species and is based on historical evidence that allows specific pruning needs to be anticipated. Programmed maintenance, which largely involves pruning but also includes other practices such as cabling and bracing, is the servicing of trees on a planned schedule, with the goal of improving and extending the function of trees and reducing the necessity of more costly maintenance later.

Programmed maintenance requires a pruning cycle, or the number of years necessary to prune all trees in a particular area. A 5-year pruning cycle, for example, simply means that one fifth of the trees will be pruned every year and that individual trees will be serviced every 5 years. Optimum pruning cycles must consider species, age, distribution, and environmental factors. In many cases, however, pruning cycles are determined less by tree needs than by budgetary and social considerations. Optimum pruning cycles may be determined by marginal cost and return analysis, identifying the relationship between pruning and tree condition. Pruning costs are measured against each annual extension of the pruning cycle and marginal returns determined. Such analysis cannot be singly applied to the total city tree population but must be done to separate situations of age, species, and other factors within the total population.

Pruning needs must be prioritized. Although hazard removal must take top priority, it is simply not enough to say that unsafe trees will be taken care of first and then others will be prevented from becoming unsafe. Prioritization requires current, accurate information on species, size, condition, location, growing environment, and past costs of maintenance. Without such information, a reasonable approach to pruning, including determination of pruning cycles and accurate projection of future budgetary needs, is not possible.

For cities just beginning comprehensive urban forest management, or those rejuvenating existing programs, pruning costs, as percentages of total program budgets, will likely be greater in earlier years because of the high priority of hazard reduction. When hazard removal declines and pruning cycles can be achieved, yearly pruning budgets (barring natural disasters) will become more stable. Annual pruning budgets can then be based on average pruning costs, which is determined by tree diameter and frequency, divided by pruning cycles. Table 6.7 shows the relationship of diameter to pruning cost. Note the substantial cost increase from 6 to 8 inches. Trees 8 inches in diameter and larger generally require climbing or special lift equipment to allow for pruning, while trees with smaller diameters do not require these more expensive techniques to accomplish the task.

Managing Pruning. Pruning may be accomplished by in-service crews, by adjacent property owners (as is required by ordinance in some cities), or by contracting with commercial arboricultural firms. In the case of pruning by property owners, the management role of the city forestry departments or tree boards is mostly educational, directed toward helping property owners do a better job or improving the quality of work of arborists whom owners engage. Management may also include notifying property owners of pruning needs of trees.

Decisions whether to prune with in-service crews or to contract with arboricultural firms are based largely on economics, although convenience, timing, and other factors must be considered. Total costs of in-service crews must be weighed against contract costs in context with the magnitude of long-range management, planned cycles for pruning, and personnel needs for other activities. Annual in-service costs include salaries, benefits, training, liability insurance, amortized equipment costs, and supplies. The total of such costs allocated to pruning, divided by the number of trees to be pruned, gives an average cost per tree. This figure provides a basis of comparison with contract quotations. When all costs are carefully analyzed, many cities find contracting to be the preferable alternative.

Bidding and Contracting. Contracting for tree pruning requires careful management. Competitive bids must be sought and contracts entered. Work must be monitored, and necessary adjustments must be made. A major difficulty in preparing for contracting is describing the work to be done. It is simply not practical to describe every cut to be made on every tree, yet those who are submitting bids must have a clear understanding of what is required. Hence, pruning bid invitations

TABLE 6.7. *Determination of Yearly Pruning Cost (5-year pruning cycle)*[a]

Diameter (Dbh)	Number of Trees	Cost Per Tree ($)	Total Cost ($)
2	225	12	2,700
4	326	14	4,564
6	365	17	6,205
8	393	26	10,218
10	349	30	10,470
12	330	35	11,550
14	310	39	12,090
16	280	43	12,040
18	295	48	14,160
20	210	54	11,340
22	160	59	9,440
24	162	64	10,368
26	95	70	6,650
28	66	77	5,082
30	62	83	5,146
32	55	91	5,005
34	21	98	2,058
36	14	105	1,470
		Total	140,556

[a]Annual pruning cost = total cost divided by pruning cycle, or $28,111 per year for 5-year cycle.

should be written in a manner most helpful to potential contractors, including descriptions of areas to be pruned, species frequency, diameter classes, and pruning standards to be followed. Pruning standards, useful in contracting, have been established by the National Arborists Association. The NAA standards recognize four classes of pruning, as summarized here:

Class I—Fine pruning. Removal of dead, dying, diseased, interfering, objectionable, and weak branches on the main trunk and within the crown, allowing for an occasional branch up to 0.5 inch in diameter if not practical to remove.

Class II—Medium pruning. Same as Class I, except for allowance for an occasional branch up to 1 inch in diameter if not practical to remove.

Class III—Coarse or safety pruning. Removal of all dead, diseased, or weak branches greater than 2 inches in diameter.

Class IV—Cutting back, topping, or drop crotch pruning. Reduction of tops, sides, underbranches, or individual limbs in situations of utility lines, unusual growth, excessive size, or severe root loss.

Although the NAA standards serve as a basis for common understanding, it is important to recognize that allowances for subjectivity must be made, as in the cases of "objectionable" and "weak" branches. Such determinations must be based on professional judgment. Therefore, it is important to ensure that qualified professionals are ultimately engaged to do pruning. Required qualifications must be clearly stated in bid invitations and may include licensing, certification, professional affiliations, or other credentialing. Care must be exercised, however, to avoid social discrimination. Figure 6.8 lists considerations in inviting bids and entering contracts for tree pruning. As is required in most cities, all bid invitations and contracts should receive careful scrutiny by the City Attorney.

As noted in item 15 of Figure 6.8, contractors may be required to furnish periodic (preferably daily) accomplishment reports, listing trees pruned by specific locations. Whether by contractors or in-service crews, such reporting is important to maintaining a continuous inventory and can be extremely valuable in case of liability claims. Ensuring timely feedback is an important part of pruning management.

MANAGING REMOVAL AND UTILIZATION

Trees, as a natural part of their life process, are always discarding something—leaves, fruit, dead branches, and the like. Trees are also mortal and ultimately must die. Such natural processes are influenced by nature (storms, drought, insects, diseases, other) and by people. In natural forests, discarding and dying is a normal part of nature's scheme. In the urban forest, though, where human society dominates, many of the things discarded by trees, and the dead bodies of trees themselves, must be removed. Some of what is removed can be used. In some instances parts of the urban forest may be harvested for fuelwood, Christmas trees, and other products. Such removals are minor, however, when we consider everything that must be removed from the urban forest including

Dead trees
Hazardous trees
Oversize or competing trees
Trees in the way of construction

1. Names, addresses, and telephone and fax numbers of involved parties.
2. Number of trees.
3. Description of trees (species, size, other).
4. Location of trees.
5. Statement of title to, or authority for, trees.
6. Description of work to be done. May include requirements for removal of dead, broken, crossing, and redundant branches; drop crotching; elimination of stubs; traffic and view clearance; wound treatment, if any; and debris disposal. Specify that all work must conform to the National Arborist Association Standards for Arboricultural Work and the American Standards Institute Safety Requirements for Tree Care.
7. Tree damage. Specify that no climbing irons, spurs, or spikes are to be used. Any damage done to trees will be immediately repaired or compensated for by the contractor.
8. Sanitation. Provision for sanitizing pruning tools or taking other measures in cases of infectious diseases or transmittable undesirable insects or other organisms.
9. Property damage. Damage done by contractor to any person or property, public or private, is the total responsibility of the contractor.
10. Public relations. May specify that contractor informs property owners or residents at least 24 hours before work is to be done and that contractor coordinates traffic control with city police.
11. Clean up. Completed within 2 hours after each pruning. Contractor may not leave site until clean up is done.
12. Contract price.
13. Time and method of payment.
14. Inspection of work and conditions for follow-up in case contract specifications are not met.
15. Reporting. Requirement that contractor furnish periodic reports of work accomplished—trees pruned and precise locations.
16. Termination. Conditions and procedures for contract termination in case of default.
17. Requirements of contractor. Requirements for licensing, certification, other credentialing, insurance, bonding, permits, or other provisions.
18. Ingress and egress rights.
19. Dates of bid submission, job completion, and contract entry.
20. Signatures of responsible parties.

Please Note: Most cities have standard bid invitation and contract forms upon which the preceding elements may be entered.

FIGURE 6.8. Considerations in inviting bids and entering contracts for tree pruning.

Stumps and sometimes roots
Hazardous branches
Other branches removed during pruning
Leaves
Obnoxious or hazardous fruit

Seeds and seed carriers, and even
Flower petals

Removal of the foregoing objects is necessary because they are either hazardous, offensive, or obstructive. Priorities for removal depend on the proximity of people and structures, with concerns for safety and sensitivity. A major key to management of removal and utilization is anticipation of removal needs. Obviously, many of these materials cannot be removed before they happen—before leaves or fruit drop, for example—but most (barring storms or other calamities) can be anticipated, and appropriate action can be taken.

Anticipating and scheduling removal is preliminary to management and is tied directly to knowledge of the urban forest gained from inventory information, operational plans, and personal observations. Inventory information reveals the tree removal and pruning needs scheduled in operational plans. It also reveals species in locations where production of fruit or other materials may cause hazards or be offensive. Personal observations provide knowledge of timing of fruit production, leaf drop, and other natural processes. Work priorities for removal can then be based on timing information and proximity of people. Proximity of people relates to hazard (target) potential and, from a management perspective, means simply that areas of high-people usage should receive high priority.

The fundamental question concerning removal is: What is to be done with the material? Can it be used or must it be disposed of in some other manner? Another question that occurs is: Can the material be used profitably—to help defray the cost of removal? In general, opportunities for profitable utilization are limited—because of volume, inconsistency of supply, handling and processing costs, and foreign materials (metal and other objects in logs, for example). The following summary of utilization possibilities lists various materials from the urban forest.

- *Branches and trunks.* Includes materials removed in maintenance pruning, hazard reduction, and whole tree removal.
  Branches are most often chipped on site and used as mulch, but they may also be processed for fuelwood.
  Trunks have a limited potential as lumber or other wood products. They often contain metal or other foreign materials. An exception may be high-value species such as black walnut and/or trees removed from natural forests during construction. Trunks can also be used for fuelwood but often must be disposed of in land-fill.
- *Stumps.* Used to be grubbed out or burned but now are ground by

machine. Residue is finer texture than chips but may be used as mulch or soil amendment.

- *Leaves.* Used to be burned but now are collected and composted or disposed of in landfill.
- *Seeds and seed carriers.* Some seeds, such as acorns, may occasionally be collected and used by nurseries. Pine cones and other seed carriers have a limited use in crafts and as ornaments.

Many cities operate urban forest product utilization facilities, particularly for fuelwood and leaf composting. In most instances, however, materials are processed by private firms, often by contract with the city. Removal is a big factor in urban forest management. It is costly and often disruptive and noisy. It frequently requires coordination with utility, police, and other units of city government. There are also legal considerations, because permits for disposal must be obtained and liability protection must be ensured. Removal and utilization management must also operate in an environment of social sensitivity. In addition to noise and disruption, people are sensitive to trees being cut down, particularly large, old specimens, even though they are hazardous. Such attitudes strongly suggest the need for effective public relations, explaining the necessity of such work in context with long-range objectives and gaining citizen understanding and tolerance.

SPECIAL CONSIDERATIONS

The preceding part of this chapter has been concerned with the primary activities of city forestry departments that are necessary to meet the fundamental establishment, maintenance, protection, and removal needs of the urban forest. What has been discussed constitutes a major part of day-to-day direct management within the forest. There are additional considerations, however, that, although not normally translating into daily activities, are important to comprehensive management.

Trees and Construction

Urban forest managers have an interest in construction as it influences existing trees and the establishment of new trees. Of particular concern are activities such as street widening, trenching for utilities, and new construction in natural forests. Although such activities may cause wounds to trunks and branches, the greater concern is normally for the rootzones of trees. Rootzone problems can result from mechanical severing of roots, soil compaction, changes in grade, and alterations to water tables. (See Figures 6.9 and 6.10.)

From a management perspective, responsibility and authority must first be considered. As previously discussed, city forestry departments generally have responsibility and authority for trees along streets, in boulevards, in city parks, and in other public areas. Hence, direct involvement is called for in construction planning and implementation in these situations. Involvement may be much more limited, however, concerning construction in natural forests. Development in natural forests is most often on private lands, and involvement of the city forester may be only advisory in the planning and approval process or regulatory during construction as authorized by ordinance.

In both situations, involvement in construction planning is critical; the interests of trees can best be represented during planning. Unless influenced by citizen action or by economic considerations, construction is often a fait accompli, and the only hope is that existing trees will be damaged to the minimum degree or that appropriate new trees may be established. Such does not need to be the case, however, and there are several things that might be done depending on the strength of public opinion or in consideration of costs and benefits.

FIGURE 6.9. Construction activity can have severe negative impacts on rootzones of trees.

FIGURE 6.10. Disregard for trees during construction can be negative to the landscape and to property values.

- Abandonment of construction plans. This response is normally applicable only in unusual cases, where historic or other special trees are involved or in special ecosystems for rare or endangered species.
- Relocation of existing utilities, sidewalks, or other structures to protect trees.
- Planned location of streets, drives, buildings, and utilities. This response requires involvement in early site planning and presents the maximum opportunity for protection. Cost/benefit considerations are critical in site planning.
- Alternative construction activity sites. Designed primarily to prevent rootzone damage, this response includes such measures as designating equipment lanes, parking areas, concentration yards, and debris disposal areas.
- Alternative construction methods. Examples are tunneling under rootzones rather than trenching, pumping concrete, and using cranes or helicopters to "drop in" materials and equipment to avoid rootzone disturbance.
- Preconstruction arboricultural practices. This response may in-

clude clearance pruning to avoid mechanical damage and fencing to protect critical rootzones.
- Postconstruction arboricultural practices. Designed to help trees adjust to environmental changes or to repair damage caused by construction, such practices as soil aeration, fertilization, and pruning may be appropriate.

Because of the efforts of urban forestry advocates, land developers and building contractors are becoming increasingly active in protecting trees and other vegetation. Often in direct proportion to opportunities for passing protection costs along to buyers, the increasing interests of developers are reflected in their engagement of private urban foresters as consultants or by improved practices resulting from workshops and seminars. One such consultant (Stewart, 1994) identifies three stages in the protection process: planning, construction, and maintenance. The three stages involve the following general activities.

- *Planning (educating ourselves)*
  Assemble the team
  Understand the site
  Identify concerns
  Analyze the site (gather data)
  Draft the preliminary plan (what to do with the facts)
  Formalize the final plan (design development)
- *Construction*
  Preconstruction activities—Preparing construction documents, selecting contractors and subcontractors, attending the preconstruction meeting, preparing the early action plan
  Construction activities (specific techniques to save trees, monitoring the process)
- *Maintenance (keeping it green)*
  Restoration and repair
  Comprehensive maintenance plan

The role of the city forester in the total process of protecting trees during construction should be clearly identified. As with all comprehensive management, involvement will be both direct and indirect. If unprescribed or limited by ordinance, the direct role—in planning, technical assistance, and enforcement—should be clarified. There should also be a strong educational component, such as arranging for workshops and distributing how-to information to developers, builders, and property owners.

In addition to protecting individual trees, attention in urban development must be given to watershed values, wetlands, critical wildlife habitat, and protection of threatened or endangered species. As related to the urban forest manager's role in overall city planning, three basic questions concerning development should be considered:

1. How may the values of natural forests scheduled for development be protected and enhanced?
2. Are such values recognized and provided for by city planners, administrators, developers, and others?
3. If such values are not recognized, what measures are necessary to ensure them?

These questions suggest the importance of the urban forest manager's involvement in comprehensive city planning and in the development process.

Fire Protection

Except in the arid west in areas where urban development has interfaced with natural forests, fire protection is of relatively minor concern to urban forest managers. Fires can occur in urban forests anywhere, however, and sometimes with disastrous results. The primary focus of urban fire protection is on structures, and forests, if considered, are generally looked upon as "carriers" of fire, which endanger buildings and other structures. Because structures receive high priority for fire protection, however, the urban forest usually benefits. Fire protection in urban forests may involve fuel reduction within natural forests, removal of fuels near structures, maintenance of fire breaks, and prevention of hazards by carefully selecting "fire-resistant" trees and other vegetation.

Fire protection in the urban forest is usually the responsibility of city, county, or other local fire departments, with state forestry departments sometimes having a role in the urban–rural interface. The urban forest manager's role, if not direct, should be cooperative with such agencies, providing input in fire protection planning and assisting in fire disaster recovery.

Urban Wildlife

Although generally highly valued in urban areas, wildlife receive relatively little consideration in day-to-day urban forest management. This statement is not meant to suggest that urban forest managers are insensitive to wildlife needs. It simply means that most management activities are driven by other needs. In fact, some management practices, such as removal of dead branches and hazard trees, may be detrimental to wildlife.

When we consider the overall physical environment of the urban forest, as discussed in Chapter 3, it is no surprise that wildlife abounds—even in greater densities for some species than in natural habitats. (See Figure 6.11.) Advantages and disadvantages of the urban environment to wildlife are shown in Figures 6.12 and 6.13. That people value urban wildlife is attested to by expenditures of about $1 billion annually for bird seed. The additional amounts spent for feeders, binoculars, books, and other bird-related items would perhaps double this figure. Urban wildlife give people enjoyment and provide city dwellers with a link to the natural world, which is of particular value in teaching children about nature. There are negatives, however, as urban wildlife can cause automobile and even aircraft accidents, may carry parasites and diseases, may be damaging to garden and other landscape plants, and may become nuisances because of their presence.

Urban wildlife management has three general aspects: providing for increased or diverse populations, providing for viewing or consumptive opportunities (the latter most applicable in the case of fisheries), and controlling damage. Some suggested wildlife management opportunities for urban forest managers follow:

FIGURE 6.11. Wildlife abounds in urban areas, often in greater numbers than in natural habitats.

- A great variety of planted trees, shrubs, and annual plants.
- Buildings of all sizes and configurations
- Wires everywhere
- Lights to attract insects
- Storm sewers for travel lanes
- Areas of natural forests
- Open parks and meadows
- Ponds, lakes, rivers, and seashores
- People who feed birds and animals, build birdhouses, and spill or discard food
- No hunting

FIGURE 6.12. What the urban environment offers wildlife.

- Large populations of cats and dogs
- Large populations of cars and trucks
- Toxic chemicals
- Wires and towers
- Buildings with reflective sides
- Glass windows

FIGURE 6.13. Urban problems for wildlife.

- Provide for travel lanes between naturally forested areas, using such features as abandoned railways, riparian strips, and other vacant lands.
- Protect wildlife habitat in new developments—meadowlands, woodlands, wetlands, and special vegetation.
- Encourage or facilitate establishment of wildlife-friendly trees and other plants.
- Leave dead and dying trees as nesting sites and insect food sources for birds in areas where such trees are not hazards.
- Erect artificial nesting structures.
- Provide access to wildlife by constructing trails, viewing platforms, boat ramps, or other structures.

Urban wildlife damage management becomes necessary when wildlife problems become intolerable. For damage management to be effective, it is necessary to understand the principles and methods. It is also necessary to recognize that management must often be applied in an environment of extreme public sensitivity to animals and that methods tolerable to some will be intolerable to others. Principles and methods of damage management follow:

- *Principles*
  Recognize damage patterns and species
  Know animal's biology and life requirements
  Know control alternatives
  Know local, state, and federal laws
  Choose control method(s)
- *Methods*
  Establish exclusion—Fencing, protective devices, construction techniques
  Lessen attraction—Habitat modification, repellents, frightening devices, resistant crops, Introduction and encouragement of predators
  Reduce population—trapping, shooting, inhibit reproduction, toxicants

Assistance with wildlife management is generally available from state wildlife departments, the Cooperative Extension Service, and the USDI Fish and Wildlife Service. An additional source of detailed information is the National Institute for Urban Wildlife, 10921 Trotting Ridge Way, Columbia, MD 21044.

Valuation

Urban forest managers are occasionally called upon to help establish values for individual trees or for urban woodlands. The need for establishing values is generally because of casualty losses or condemnation of properties to be developed. Traditionally, woody plants in the landscape have had no real value in themselves; their only value was their influence on the value of the real estate where they are (or were) located. Change is being advocated, however, and some states are considering legislation to allow woody plants to be valued unto themselves.

There are various methods for assigning monetary values to woody plants in the urban forest. The appropriate method depends on plant size, species, condition, function, location in the landscape, and other situational factors. Because of these factors, particularly location in the landscape, valuation can be subjective. Thus, the urban forest manager can offer only his or her best professional judgment as to the value of a particular woody plant. Methods of appraising values and their suggested applicability follow:

Direct replacement cost. Cost of replacing with same species of near-identical size, shape, condition, and other characteristics. Generally applicable to small trees and shrubs.

Compounded replacement cost. Cost of establishing transplantable-size plant of same species compounded for number of years to reach size of plant being replaced. Generally applicable to medium size trees and large shrubs.

Present value of future returns. Value at economic maturity discounted to time of loss. Most applicable to short rotation crop trees such as Christmas trees.

Cost of repair. Cost of arboricultural treatment to provide for restoration of plant. Applicable in cases of repairable damage.

Cost of cure. Cost of treatment necessary to return plants to a reasonable level of original condition by correcting conditions responsible for damage. Applicable in situations where causal agents are clearly identifiable and correctable.

Forest value. Generally based on lumber, veneer, fuelwood, or other wood product values but can apply to other values such as wildlife and watershed. Most applicable to natural woodlands or forest plantings within the urban forest. Could include landscape trees such as black walnut, black cherry, and other high-value species if of merchantable quality.

Crop value. Value of annual crops over the remaining lifespan of the species. Most applicable to fruit or nut trees.

Trunk formula. Used for trees too large to practically replace. Based on cost per square inch of trunk area as determined by replacement cost of largest locally available transplantable tree and adjusted by species, condition, and location factors.

The trunk formula method, generally applicable to large trees, is a recent adaptation of the valuation formula of the International Society of Arboriculture (ISA), which was based on a predetermined value per square inch of trunk diameter (periodically adjusted for inflation). This value, applicable nationwide, was then adjusted by species, condition, and location factors. The latest formula, developed in 1992 by the Council of Tree and Landscape Appraisers, replaces the ISA base value with a base value derived from the cost per square inch of trunk diameter of the largest locally available transplantable tree of the same species of the tree being appraised. This value is applied to the total square inch trunk area of the appraisal tree and then reduced according to species class, condition (structural integrity, health), and location (site, contribution, placement) in the landscape. Figure 6.14 is a worksheet with guides for determining appraised value (International Society of Arboriculture, 1992).

A detailed treatment of valuation methods is given in *Guide for Plant Appraisal,* Eighth Edition, 1992, by the Council of Tree and Landscape

TRUNK FORMULA METHOD FORM

Appraised Value = Basic Value x Condition % x Location %
Basic Value = Replacement Cost + (Basic Price x [TA$_A$ - TA$_R$] x Species %)

1. **Replacement Cost:** largest transplantable tree* $ _____ . ____

2. **Basic Price** of replacement tree* $ _____ /in²(cm²)

3. Difference in trunk areas of
   appraised & replacement trees
   A. Appraised tree trunk area (**TA$_A$** or **ATA$_A$**)**   _____ in²
   B. Replacement tree trunk area (ᵀᴬ•)*   _____ in²

   C. Difference in trunk areas   _____ in²

4. Multiply **Basic Price** difference in trunk areas
   $ _____ /in²(cm²) x _____ in²(cm²)   = $ _____ . ____
     (Line 2)   (Line 3C)

5. Adjust Line 4 by **Species** rating* ____%   = $ _____ . ____

6. **Basic Value** = $ _____ . ____ + $ _____ . ____   = $ _____ . ____
                     (Line 1)   (Line 5)

7. Adjust Line 6 by **Condition** ____% = $ _____ . ____

8. Adjust Line 7 for **Location:**
   Location = (Site + Contribution + Placement) ÷ 3
         = (____% + ____% + ____%) ÷ 3 =   ____% = $ _____ . ____

9. **Appraised Value** = Round Line 8 to nearest $100   = $ _____ . ____

FIGURE 6.14. Worksheet for determining appraised value—trunk formula method.

Appraisers, and published by the International Society of Arboriculture, P.O. Box GG, Savoy, IL 61874.

INFORMATION MANAGEMENT

This chapter ends by emphasizing the importance of record keeping and information management to effective urban forest management. Keeping accurate records and being able to retrieve information is absolutely essential. Records are not simply historical documents but are invaluable tools for future management. Records reveal what has been done to trees and serve as the basis for determining future needs. Records allow evaluations to be made and are essential for planning and budgeting. Records also furnish proof in cases of disputes involving liability. Modern computer technology allows massive amounts of information to be efficiently handled. The weakness, all too often, is that information is not currently entered. It could be argued that information entry is the most important part of urban forest management, for without accurate information there would be no basis for making reasonable decisions.

Information management should not be limited to internal records. Technology is readily at hand to access information on virtually any topic relevant to urban forestry. Urban forest managers can tap various computer networks and can communicate with contemporaries in other cities. Computer programs are available for tree inventories, budgeting, cost accounting, program analysis, tree selection and location, and other management applications. All are tools in helping urban foresters make the best possible management decisions.

SUMMARY

The objectives of management are to meet the establishment, mainte-nance, protection, and removal (including utilization) needs of the urban forest. Management of tree establishment must consider the various factors of location and selection, the application of resources to planting, provisions for ensuring survival and health, and evaluation of the results. There are three basic, but interrelated, areas of mainte-nance management: (1) hazard management, (2) plant health manage-ment, and (3) quality improvement. Pruning is a major factor in each. Protection involves plant health management, providing for trees dur-ing construction, and protection from fire. Removal of vegetative mate-rial is a costly necessity requiring sensitivity to the needs and desires of occupants of the urban forest. Urban wildlife management has three aspects: (1) providing for increased or diverse populations, (2) providing for contact opportunities, and (3) controlling damage. Monetary values of trees are assigned by a variety of methods, depending on plant size, species, condition, location, function, and other situational factors. Record keeping and information management are extremely important. Records reveal what has been done to trees, serve as the basis for determining future needs, allow evaluations to be made, and furnish proof in cases of liability.

REFERENCES

Aber, J. D. 1990. "Forest Ecology and the Forest Ecosystem." *Introduction to Forest Science,* Second Edition, edited by R. A. Young and R. L. Giese, pp. 122–23. John Wiley & Sons, Inc., New York.

Ball, J. 1994. Plant Health Care and the Public. *Journal of Arboriculture,* 20(1): 33.

Grey, G. W. 1993. *A Handbook for Tree Board Members,* p. 21. The National Arbor Day Foundation, Lincoln, NE.

Grey, G. W., and F. J. Deneke, 1986. *Urban Forestry,* Second Edition, p. 139. John Wiley & Sons, Inc., New York.

Harris, R. W. 1983. *Arboriculture: Care of Trees, Shrubs, and Vines in the Landscape,* p. 50. Prentice-Hall, Inc., Englewood Cliffs, NJ.

International Society of Arboriculture. 1992. *Workbook, Guide for Plant Appraisal,* Eighth Edition, p. 10. International Society of Arboriculture, Savoy, IL.

Loomis, R. C., and W. H. Padgett. 1975. *Air Pollution and Trees in the East.* USDA Forest Service, Upper Darby, PA.

Lumis, G. P., G. Hofstra, and R. Hall. 1973. Sensitivity of Roadside Trees and Shrubs to Aerial Drift of Deicing Salts. *Horticultural Science,* 8(6):475–77.

Miller, R. W. 1988. *Urban Forestry: Planning and Managing Urban Greenspaces,* Prentice-Hall, Inc., Englewood Cliffs, NJ, p. 191.

Pirone, P. P., et al. *Tree Maintenance,* Sixth Edition, Oxford University Press, New York, NY, p. 211, 1988.

Society of American Foresters. December 1994. "SAF Forest Entomology and Pathology Working Group Newsletter (D-5)." Society of American Foresters, Bethesda, MD.

Stewart, C. January 21, 1994. "Building With Trees," Program from workshop held in Savannah, GA.

Whitlow, T. H., and R. W. Harris. 1979. "Flood Tolerance in Plants: A State-of-the-Art Review." U. S. Army Engineer Waterways Exp. Station Technical Report E-79-2, Vicksburg, MS.

Yamamoto, S. T. 1985. Programmed Tree Pruning and Public Liability. *Journal of Arboriculture* 11(1):17.

CHAPTER 7

Leveraging Your Efforts

The preceding chapters have dealt with the application of management—how the needs of the urban forest are met through planning, budgeting, and implementation. In this chapter we look at ways of gaining support, extending management, and improving the quality of services to the urban forest. The topics addressed are directed toward three objectives vital to comprehensive management: a strong city forestry department, effective use of external resources, and high-quality performance from those who service the urban forest. Specific topics include identifying program needs, building program support, extending management, working with the media, working with volunteers, and securing financial and technical help.

IDENTIFYING PROGRAM NEEDS

Identification of urban forestry program needs is a product of analysis and should begin with the fundamental question: Is the program as effective as it should be in meeting the needs of the urban forest? Properly, the answer should be "No," as all programs can be improved. There is the possibility, though, that improvements in effective programs can provide marginal returns. It is my judgment, however, that such situations are extremely rare.

Program analysis must be made on both the macro and micro levels.

Macro analysis is an administrative function and requires a look at the six elements necessary for comprehensive management: a central organization with responsibility and authority, knowledge of the total urban forestry environment, knowledge of what the urban forest needs, plans for meeting the needs, adequate budgets, and effective implementation. If any element is completely missing, comprehensive management cannot occur. If any element is partially missing, management cannot be as effective as it could be. Once missing elements are identified, additional questions must be asked: Why is the element missing? What are the barriers to putting it in place? and How may the barriers be overcome? The answers identify what should be the highest priorities of urban forestry program administrators.

In most macro analyses of urban forestry programs, adequate budgets will be identified as the major missing element. This is an easy call, because most managers do not feel that their budgets allow them to do all that they want to do. The adequacy of budgets must be determined, however, in relation to long-term objectives. If the program is "on target" in meeting the objectives as planned, budgets are probably adequate. Adequate budgets are, of course, frequently missing and may also influence other necessary elements for comprehensive management. For example, budgets may not allow for inventories to determine what the urban forest needs; and inadequate budgets obviously cannot provide for effective implementation. Program strength is tied directly to budgets, and budgets are ultimate products of citizen interest in the urban forest. Gaining citizen support is considered in the next section.

Micro program analysis involves looking within each of the six elements to identify strengths and weaknesses, and must be ongoing. While identification of deficiencies is inherent in program analysis and is most often the main focus, strengths must also be considered. Reduced to its simplest form, the basic questions in micro analysis involve what is wrong and how it might be fixed and what works well and whether more of it should be done. For each element necessary to comprehensive management, the following should be considered.

- *A central organization with responsibility and authority*
  Is authority commensurate with responsibility?
  Are responsibilities clearly defined?
  Are ordinances adequate?
  Is there a citizens' tree board or similar organization? If so, is its role clearly defined and understood?
- *Knowledge of the total urban forestry environment*
  Is the concept of the total urban area as a forest understood?
  Are ownerships and those responsible for management identified?

Is the departmental position within city government and its relationship to other functions of government clearly defined?

Are other players in urban forestry identified—media, support organizations, arboricultural and nursery firms, and others?

Is the political environment understood?

- *Knowledge of what the urban forest needs*
Is there current survey or inventory information?
- *Plans for meeting the needs*
Is there a long-range plan with clearly defined objectives and priorities?

Is there a current operational (annual) plan?

Is the operational plan consistent with the objectives of the long-range plan?

Is there a current plan of work?

Is there planned latitude for unforeseen situations short of major disasters?

Is there a disaster emergency plan?

- *Adequate budgets*
Are budgets adequate to address the objectives of the long-range plan successfully?

Are budgets derived from an operational plan rather than adjusted for inflation?

- *Effective implementation*
Are all management operations interrelated and based on objectives of the long-range plan?

Are all management operations consistent with operational plans?

Is there an effective and efficient information management system?

Although closely related, program needs analysis differs from operations analysis as discussed in Chapter 6. Program needs analysis deals with what is necessary for management to take place. Operations analysis concerns specific functions of management.

BUILDING PROGRAM SUPPORT

Without exception, all successful city forestry programs that I have observed have strong citizen support, expressing to city administrators that trees should receive high priority. It has been my further observation that each city having a successful program also has key individuals who are strong advocates for trees. A final ingredient of each successful program is a city forester with a strong sense of political and public relations leadership—one who recognizes that urban forestry must compete with other needs of the city and sees orchestration of public support as a vital part of day-to-day management.

A fact fundamental to success, and sometimes lost on urban forest managers, is that city administrators (mayors, council members, others) are responsive to citizen wishes. Another fact is that administrators are particularly responsive to wishes expressed by citizens in positions of power. Hence, it follows that people in positions of power who are advocates of trees can have a strong influence on urban forestry programs. The obvious suggestion from this observation is that the city forester should cultivate an advocacy relationship with powerful citizens. Nevertheless, some city foresters may find such political activities distasteful or, lacking confidence, will refrain from cultivating such relationships. A final fact, however, is that city foresters who are not politically active are most often in charge of programs lacking the resources to be successful.

The preceding reference to power is not meant to suggest that powerful people are representatives of the underworld or are ruthless in protecting their own interests. Quite the opposite is true. Powerful people are those who are respected in their own right and represent broad community interests. They may be merchants, professional people, or homemakers who simply like trees or want their own neighborhoods to be made better because of trees. They may be representatives of garden clubs or other organizations with interests in trees. They may be business owners or executives with a sincere interest in improving the environment of their city. And they may be representatives of youth clubs or other organizations who see involvement with trees as a means toward other social agendas.

In addition to the obvious implications for program budgets, involving key people can be a factor in program implementation, because many of these people will become volunteers for tree planting and other activities. Working with volunteers is discussed in a later section.

Building program support can be related to marketing, and it is useful for urban forest managers to think in such terms. The product being marketed is a better urban forest environment, and the purchase price is citizen interest as expressed by key individuals. Marketing to key individuals involves such things as asking them to serve as planning advisors, recommending them for appointments to advisory committees, giving talks or presenting programs to groups they represent, and appealing to their motivations for being interested in trees.

Tree boards can be extremely important in building program support. (Please refer again to Appendix 1.) Committed to improving the urban forest environment, individual board members can greatly extend the effectiveness of the city forester in cultivating support. Few things are perhaps more important than securing appointments of the best possible people to tree boards. Ideally, tree board members should themselves be powerful individuals, as described previously. They should see their advocacy role clearly, in context with total long-range program needs and the city's particular political and social environment. Much advocacy will occur during day-to-day business and social activities of individual board members, and some citizen interest will result from normal tree board operations. There should, however, be a planned program of public relations to keep trees in the forefront.

Keeping Trees in the Forefront

A discussion of public relations in urban forestry should begin with the understanding that public relations—good, bad, or indifferent—exist simply because an urban forestry program exists. To paraphrase a recent popular bumper sticker: "Public Relations Happens." All organizations and individuals have public relations simply because they are there, and the product is variously known as image, perception, or reputation. Images, perceptions, or reputations result from normal social and commercial intercourse but can be deliberately influenced (hopefully for the better) by planned activities, normally with the objective of helping individuals or organizations get along better in the world. In urban forestry, the objective of a favorable (and accurate) image is strong program support enabling the needs of the forest to be met.

As suggested, urban forestry public relations must be ongoing, through day-to-day activities sensitive to the public, and should also be planned. Public relations planning in urban forestry involves the following principles and process.

- *Principles*
  Keep trees, rather than urban forests, in the forefront. People relate to trees but do not consider themselves as living in a forest.
  Identify specific audiences.
  Use methods appropriate to audiences.
  Identify teachable moments.
  Seek and act upon every reasonable opportunity.
- *Process*
  Identify public relations opportunities in scheduled program activities.
  Plan and create public relations events.
  Take advantage of teachable moments.

The first step in public relations planning for urban forestry is to look at scheduled activities and identify opportunities for keeping trees in the forefront. Scheduled activities may include inventorying, hazard tree inspections, tree planting, pruning, health management, or other operations. For each program activity there will be three audiences: observers, those who are directly affected, and those in whom interest might be created. Observers are those who actually see the activity and may be strongly or idly curious as to what is going on. Those who are directly affected are usually adjacent property owners or residents who may be temporarily inconvenienced but may receive long-term benefits from the work. Those in whom interest might be created may be frequent users of the affected portion of the forest, other indirect stakeholders, or even individuals who might be placed in the elusive and perhaps inaccurate category of the "general public."

Methods must be appropriate to audiences. For observers, signs explaining the work (even as basic as "Tree pruning in progress for the health and safety of your trees. Thank you for your patience.") can be helpful. Doorhangers and other flyers can also be used. Most effective, however, if practical, are personal explanations by qualified and artic- ulate departmental personnel. These methods, plus direct mail, are also appropriate for those who will be directly affected. Reaching others in whom interest might be created usually involves using the media. Working with the media is discussed in a later section.

The second step in urban forestry public relations planning is the creation of events to keep trees in the forefront. Such events—in conjunc- tion with other occasions, such as parades, festivals, or other ceremo- nies—may be related to holidays or observances or to celebrations of trees themselves such as Arbor Day observances, urban forestry confer- ences, or tree appreciation banquets. In each celebration the focus should be on trees and people, not on urban forestry programs. Such events provide excellent photo opportunities for city administrators and offer recognition to volunteers and others who serve the community through their work with the urban forest. The events also provide the media with human interest stories, positively received by their audiences.

The concept of the teachable moment is based on the fact that most people are not interested in a particular topic unless they have a need to be interested. The need may be because of an emergency or other occurrence that sparks their interest. Teachable moments in urban forestry occur when natural or contrived events draw attention to trees. A sampling of these events include

A storm, fire, or other disaster
Insect or disease outbreaks

Special trees blossoming in the spring
Leaves changing color in the fall
Leaves falling and needing to be raked
Unusually cold, hot, dry, or wet weather that threatens trees
Tree planting time
Vandalism or other damage to heritage or other special trees
Controversies regarding trees
Arbor Day
Other events or celebrations relating to trees, e.g. Tree City U.S.A
    awards
Maintenance work on trees that directly affect people

A valuable resource for urban forest managers is *Public Relations and Communications for Natural Resource Managers,* Second Edition, by James R. Fazio and Douglas L. Gilbert (Kendall/Hunt Publishing Company, Dubuque, IA, 1988).

EXTENDING MANAGEMENT

In Chapter 1 we explored the concept of direct and indirect management. Direct management relates to areas of the urban forest for which the city forestry department has responsibility and authority for treatment. Indirect management applies to all other parts of the forest, largely in private ownership, where care can only be influenced (except for matters of public health and safety or other reasons as provided by ordinance). Indirect management of these portions of the urban forest can be accomplished by improving the knowledge and skills of owners and managers, improving the knowledge and skills of those who provide services, and improving the quality and variety of plant materials available to owners and managers.

Disseminating Information

The objective of information dissemination is education, in the hope that new knowledge will be applied to the urban forest. Thus, information cannot be disseminated at random; the audiences and their needs must be considered. Recall that indirect management has as its basis the judgment of a central organization (the city forestry department) of what is in the best interests of the total urban forest. This imposition of a judgment made by a central authority should not be offensive if we keep in mind the fact that exercise of such judgment, based on detailed information, is in the best interests of the urban forest and the health, safety, convenience, and well being of residents. It is a proper role of city government.

Needs of the urban forest may be generic, as identified in the long-range plan (diversity, safety , health, function), or specific, as related to storm damage, insect and disease infestations, poor arboricultural practices, or other situations. Needs, as identified, can be addressed according to the following principles and methods of information dissemination. Note that some are identical to the principles and practices of public relations discussed in the previous section.

- *Principles*
  Identify specific audiences.
  Use methods appropriate to audiences.
  Take advantage of the teachable moment.
  Be specific as to what you want the audience to do.
  Keep the message simple.
  Use others to tell your story.
  Repeat, repeat, repeat.
- *Methods*
  Public meetings—Particularly applicable in cases of emergencies or issues of strong interest to residents.

  Tours, demonstrations, and field trips—Appropriate for groups with specific interests or for "grand openings" of new facilities.

  Flyers, doorhangers, and posters—Can be distributed by volunteers to homes and businesses. Particularly useful in notifying residents of public meetings and creating a sense of urgency.

  Inserts in local newspapers—Most applicable in cases where residents are already aware of a situation (gypsy moth, for example). Provides detailed information and gives specific recommendations.

  Inserts in utility statements—Must be in cooperation with utility companies. Most often will relate to utility arboriculture. May provide a low-cost opportunity, particularly if insert does not increase postage rate.

  Newspaper special editions—Many newspapers publish special garden, conservation, or environment editions and may welcome articles about tree care.

  Telephone hot lines—Most applicable in situations of severe insect or disease outbreaks or other unusual occurrences. Special information numbers can established where callers may receive recorded or "live" information.

  News articles—See section in this chapter on working with the media.

Articles by garden editors—See section in this chapter on working with the media.

Television and radio—See section in this chapter on working with the media.

Direct mail—Cost effectiveness must be carefully considered. Generally applicable in cases where property owners are notified of specific actions required by ordinance.

Of the preceding principles, using others to tell your story is especially important, because it represents the best opportunity to extend information. "Others" can include Cooperative Extension Service agents, state forestry agency personnel, garden editors of newspapers, arborists, nursery people, and perhaps even the clergy. Of these, the Cooperative Extension Service can be of particular value, as agents are generally well informed concerning urban forestry matters and have effective networks for disseminating information. An additional method of using others to tell your story is to "train the trainers." This approach, used effectively by the Cooperative Extension Service through the Master Gardener program, provides training to key individuals who then extend their knowledge within their own neighborhoods. This method can also be applied to the needs of the urban forest.

Effectiveness of information dissemination is often difficult to measure, as the ultimate objective is to change human behavior relating to the care of the urban forest. The final measure is the quality of the urban forest, and betterment is often not discernable for several years. Information dissemination is but one approach to improving the urban forest, however, and coupled with improving the quality of arboricultural work and providing for better nursery services, much can be accomplished.

Working with Arborists

As stated in Chapter 3, perhaps the most effective way to improve the quality of the urban forest is to improve the knowledge and skills of those who physically work with trees. No one is in a better position to do good or harm to trees than the person with a saw in his or her hand. A program that I was privileged to be involved with may illustrate how important working with arborists can be.

Although the Cooperative Extension Service, the state forestry department, and the Kansas Arborists Association (KAA) had advocated proper urban tree care for many years, topping, or dehorning, trees was a common practice in cities in Kansas in the mid-1970s. Although some competent arborists refused to top trees, most did so with the rational-

ization that "If I don't do it, someone else will, and I will lose the money."
Concerned about topping and other improper tree care practices, the
Kansas Arborists Association initiated an arborist certification program in 1975. Administered by the KAA in cooperation with urban
forestry and arboricultural specialists from Kansas State University,
the program requires that certified arborists meet the following standards:

- Membership in the Kansas Arborists Association
- Successful completion of a course in arboriculture as prescribed or
  approved by the KAA
- Minimum of 2 years experience as a practicing arborist
- Practice of a code of ethics
- Payment of an annual certification fee
- Possession of property damage and personal liability insurance in
  an amount prescribed by the KAA

A key factor in the program is an intensive, 5-day training course that
covers technical, business, and public relations aspects of arboriculture.
Since its inception in 1975, 860 persons have completed the course, and
215 have become certified arborists (Nighswonger, 1994). Another key
factor is the code of ethics, which requires certified arborists to explain
the alternatives to topping to potential clients. Tree topping has also
been addressed legally, with some cities now having ordinances prohibiting the practice on public property and many requiring that only
certified arborists may contract for public tree work.

Although such legal requirements are important ramifications of the
certification program, the major benefit has resulted from the improved
work of both certified arborists and others who have completed the
training course. Once common, freshly topped trees have virtually
disappeared from the Kansas landscape. In their places are countless
examples of vigorous trees, well pruned and well maintained.

From the management perspective of city forestry departments, legal
and policy requirements concerning those who do arboricultural work
are extremely important. Of equal importance are measures to ensure
high quality of work. In addition to support of state or regional programs
with objectives similar to those described previously, there are other
measures that may be taken within individual cities.

- Obtain lists of all tree care firms and institutional groundskeepers
  in the city. Names are available from telephone directories and
  from city hall (if business permits or licenses are required).
- Arrange for pruning schools and other training sessions, using city

forestry department personnel, university specialists, or private arborists as instructors.
* Write a newsletter for arborists.
* Encourage arborists to become members of the International Society of Arboriculture, the National Arborists Association, and other professional organizations.
* Familiarize arborists with the city's long-range urban forestry plan and how it relates to their work.

Selecting an Arborist. In addition to regulatory measures and education of arborists, property owners and other managers of trees need to select competent arborists. This is a difficult area, because many people simply do not know the difference between proper and improper tree care and are often influenced to have unnecessary or incorrect work done. This is particularly evident in areas where tree topping is common. (See Figure 7.1.) Tree topping feeds upon itself; topped trees offer dramatic evidence, as properly pruned trees do not, that they have been "pruned." Topping thus becomes the thing to do, especially when promoted by an appeal to fear (Those limbs will fall on your house!). Natural events can greatly exacerbate the practice. A case in point is in southern

FIGURE 7.1. Tree topping can become so common that it appears to be the proper thing to do.

Virginia where a series of ice storms during the winter of 1993–94 inflicted severe damage on trees. These storms had the effect of vindicating tree topping, if you followed this logic: "If your trees had been topped, you wouldn't have this damage." Topping became so common after the storms that the real tragedy was not the physical damage to trees from the ice, but the damage inflicted by "arborists."

The case in southern Virginia is extreme (although not unique) and calls for a variety of educational and perhaps regulatory measures concerning arborists. Also critically needed for owners of trees are educational programs that will help them to recognize the alternatives to topping and to select competent arborists. The following recommendations to tree owners can be helpful in their selection of arborists.

- *Look for indicators of competency*
  Ads in the Yellow Pages. A well-designed ad is itself an indication of a serious business approach. Within the ad are also other indicators. Consider the following questions: Is topping advertised? Are a variety of services listed? Are professional affiliations listed? Is "fully licensed and insured" listed? Are references offered? Are free estimates offered?
- *Check the indicators*
  If topping is advertised, look for another ad.
  Many arborists offer a variety of services (pruning, fertilizing, pest control, cabling and bracing, and other). Ask specifically about their experience with what you want done.
  While membership in a professional arboricultural organization does not in itself ensure competency, it does suggest that the arborist is serious about professional improvement. Ask about his or her involvement in the state arborists' association, the International Society of Arboriculture, the National Arborist Association, or other listed organizations. Ask also if employees attend training sessions.
  Ask about personal property and liability insurance, including the name of the insurance company. Under some circumstances, property owners can be held financially responsible for damage caused by uninsured workers.
  Ask for references. Call them and then go see the results of the work done.
  While there is no obligation to accept free estimates, it should be understood that the cost of estimates is eventually reflected in the contract price.
- *Beware of door-knockers, especially those who offer a special price for a limited time.* Door-knockers are especially common after

storms. Unless a storm has left an immediate hazard or great inconvenience, trees can probably wait until a qualified arborist can be engaged.

Working with Public Utility Arborists. As has been noted, public utilities, particularly electric companies, exert a strong influence on the urban forest. (See Figure 7.2.) This influence can be negative, because the primary objective of utility arboriculture is the provision of electric or other services rather than the improvement of trees. Modern utility arboriculture stresses a positive influence on the urban forest, however, and city forestry departments can facilitate such an influence.

Utility companies sometimes have their own tree crews. More often, though, line clearance work is contracted to private tree service companies. Many utility companies apply strict standards to protect trees, and some engage in educational programs to help property owners better select and plant trees near lines. Some also help underwrite planting costs, providing that troublesome, costly existing trees can be removed. City foresters can help ensure better care of trees around utility lines by the following means.

FIGURE 7.2. Public utility companies, and the arborists they engage, exert a strong influence on the urban forest.

- Helping make property owners and residents more aware of the need for utility line clearance and the needs of trees.
- Meeting with utility company representatives to communicate concerns and seek their help.
- Asking to review line clearance specifications and pruning standards.
- Sponsoring or arranging educational and training opportunities for utility arborists.
- Encouraging participation in programs such as Tree Line USA, sponsored by The National Arbor Day Foundation to promote the goals of dependable utility service and abundant, healthy trees. Companies meeting the three criteria of the program—quality tree care, annual worker training, and tree planting and public education—are publicly recognized (The National Arbor Day Foundation, 1993).

Working with Nurseries

Nurseries and other suppliers of plant materials positively influence two important needs within the urban forest: species diversity and planting stock quality. For obvious reasons, most nurseries concentrate on popular species that sell readily. In many cases, popular species are also those with relatively low production costs. Unfortunately, popular species are commonly overplanted, resulting in urban tree populations that lack diversity. Of particular concern in recent years is the proliferation of Bradford pear and other flowering species.

Nursery people, if aware of diversity problems, are often willing to make a variety of species available and promote their use in the urban forest. A strong attempt should be made by urban forest managers to inform nursery people of such problems and ask for their help. Simply sharing inventory and other information concerning the needs of the urban forest can be helpful. On the Great Plains and in the arid west where species able to thrive without irrigation and other special care are limited, nurseries can be especially helpful in introducing and testing new species, and making them available if they are found to be adaptable. Also, nurseries can often accommodate needs for exotic species or other unusual trees for special landscaping situations. Species survival information can be of value to nursery suppliers. By sharing inventory information, annual survival data, and general observations, urban forest managers can help nursery producers and suppliers make decisions about species adaptability.

Planting stock quality can also be positively influenced by working with nurseries and other suppliers of plant materials. Tactful, on-site discussions of stock quality, conveying genuine concern for improving the urban forest can be productive. Urban forest managers are also concerned about materials sold by department stores, building materials retailers, and discount outlets. Frequently displayed on exposed asphalt or concrete lots, plant materials can deteriorate rapidly. Contacts with personnel of such outlets, offering recommendations as to shading, watering, and other care can result in better tree survival.

WORKING WITH THE MEDIA

If trees are to be kept in the forefront, the media (newspapers, television, radio) must be used. To use the media effectively, urban forest managers must first be aware that everyone else with a cause wants to use the media too and that the cause of trees must compete with them. They must also understand that the various media have profit motives, providing informational services in exchange for revenues from subscriptions and advertising, with revenues being directly related to audience or customer numbers. Informational services include news and commentary, entertainment, or general information and public service announcements. Thus, the chance of the media using something concerning trees is directly related to whether it is of value as news, entertainment, general information, or as a needed public service. Of these, a needed public service, such as what property owners should do to trees following a storm, will have the best chance, with news value presenting the second best opportunity. Overlaid is the element of "human interest," with editors and producers always asking the question: Will it be of interest to our readers, viewers, or listeners? The strongest interest will occur when external events create the desire for information. In no area of public relations is the concept of the teachable moment more important than when working with the media. The following recommendations should be reviewed when trying to get the media to use stories about trees.

- Always ask yourself: Is this story newsworthy? Does it appeal to the values or interests of the audience?
- Remember that your stories should be about trees. People relate to trees, particularly their trees, and not to the urban forest nor to your program.
- Take advantage of the teachable moment—when some external event causes people to be interested in trees.
- Get to know media people, the staff who put stories together and the editors who make decisions as to what is printed or broadcast. When such people recognize you as the local authority on trees,

they will seek you out when something newsworthy about trees
occurs.

- If you have written or visual material, deliver it personally. Remember that everyone else is sending them stuff.
- Develop your own regularly scheduled newspaper column or radio or television show. "Tree talk" columns or shows can be very successful. A notable example is in St. Louis, Missouri, where a radio talk show by an urban forester has the highest listener ratings of any show broadcast by the station (Kincaid, 1993).
- Remember that the media love a controversy. When there are disputes about trees (usually when trees are threatened to be removed), your presentation of a reasonable view will be newsworthy. Approach this with caution, as there can be many pitfalls.
- Use your mayor's news conferences to convey information about trees. This approach is most likely if your information is newsworthy and makes the mayor look good.
- Be a guest on someone else's show.
- Buy some media time. The surest way to get your story told (but not necessarily listened to) is to pay for it.
- Use public service announcements (PSAs). The media are legally bound to provide space and time for announcements in the public interest. You can produce your own PSAs or obtain them from your state forestry agency, the USDA Forest Service, The National Arbor Day Foundation, American Forests, and other organizations.
- Keep it simple.

An additional important aspect of working with the media is giving interviews. Television and radio interviews are often informal, conducted on-site or by telephone; they also may be formal, in-office visits if detailed information is desired. The Society of American Foresters (1995) has prepared tips on dealing with the media, which form the basis for the following recommendations for giving interviews.

- Be responsive. When an interview is requested, give reporter your name (spell it) and telephone numbers (office and home). Always return a call. The reporter is probably facing a deadline. Remember that you are not preparing a news story but helping the reporter with his or her story.
- Prepare a fact sheet in advance. Keep handy a list of facts to which you can refer. Give your fact sheet to the media.
- Have quotes ready. Keep a card index of good quotes. Review and choose relevant ones before a media interview or tour. Use your quotes early in the interview and only if appropriate.

- Dress appropriately. Always look professional. Your dress is an important part of professionalism. Remember that, if you are on camera, your audiences' impression will be formed both by how you look and what you say.
- Don't come on as aggressive, defensive, or regressive. Act as if you are happy to add information to the story, as is your role.
- Keep control. Don't loose sight of the message you wish to convey. Keep coming back to the message if you sense that the interviewer is trying to lead you.
- Keep the word *not* out of your conversation. The word *not* is defensive, as in the argument: "We are not doing what you have reported." Speak in positive terms.
- Referrals are fine. If you cannot speak authoritatively on the issue, refer the reporter to someone who can.
- Avoid technical terms and jargon. Technical terms such as *urban silviculture, basal area,* and even *species diversity* may be confusing to lay audiences. Technical jargon such as *urban infrastructure, urban interface,* and *urban ecosystem* may sound pompous.
- Find a local angle. The surest way to create interest in your message is to tie it to the locality and the values and interests of local people.
- Use statistics. Use statistics sparingly and wisely. Logic and facts often do not win over emotional appeals. Include statistics on your fact sheet so that reporters will have accurate figures when preparing their stories.
- Be succinct. Keep your statements short and factual. If you expound at length, you may find that your statements are excessively edited—or worse, your comments may be eliminated entirely in favor of someone who kept it short.
- Watch for signs. When reporters start writing and recording, it signals that you are probably on the right track.
- Smile. Smile, even if it is a telephone interview. People simply come across better when they smile.

WORKING WITH VOLUNTEERS

An effective way of extending the resources of city forestry departments is through volunteers. For urban forest managers, the first need concerning volunteers is to understand what tasks they might reasonably perform. Generally, volunteers can do such things as coordinate tree planting projects, plant trees, build facilities, improve wildlife habitat, remove litter, and distribute information. Depending on individual skills, they can also answer technical questions, provide instruction on tree care, teach children about trees, and coordinate and supervise other volunteers. Of these, and others, by far the most common involvement of volunteers is in tree planting, often to the point that more trees are planted than can be established. A common complaint of urban forest managers is that, although volunteers plant trees, they do not have the commitment to seeing that the trees become established. Another complaint is that volunteer activists are sometimes "loose cannons," imposing uncoordinated tree-planting projects on the city without considering the long-range objectives or the city's capability to care for the trees. Such complaints aside, volunteers provide valuable services to the urban forest, and every successful city forestry program of which I am aware uses citizen volunteers.

Urban forest managers need to consider volunteers as a management resource; they are not only a source of labor, but also a source of involved program support. The problem is that volunteers are a difficult resource to assess in advance and thus an equally difficult resource to apply to management planning. The urban forest manager simply does not know the "strength" of volunteers in advance. This difficulty can be partially overcome during annual program planning by looking carefully at each management task and determining if volunteers can reasonably be involved and then testing to see if they can be recruited. Shelf project plans can also be used. Shelf plans are for specific small-scale tree planting, landscaping, or other projects that can be accomplished easily by volunteers. Such projects should be location specific, such as a park, block, or section of a boulevard, and preferably limited enough to accomplish in a single day or less. These projects are important, but not critical in a given year, and must be consistent with the city's long-range plan for the total urban forest.

Urban forest managers also need to know the types of volunteer groups and a bit about the volunteers themselves.

About Volunteer Groups

Volunteer groups tend to be advocacy, project, or program oriented.

- *Advocacy groups*
  Focused on causes (often single issues).
  Strong emphasis on fund raising and lobbying.
- *Project-oriented groups*
  Often part of a larger organization. For example, the beautification committee of a local service club.
  Focused on individual projects, such as tree planting or other environmental improvements.

- *Program-oriented groups*
  Focused on broader, generally long-range concerns. For example, tree boards with interests in the overall urban forest.

About Volunteers in General

Volunteers are a special kind of people.

- Volunteers will come forth if there is a sense of urgency.
- Volunteers need to be asked.
- Volunteers see what they do as a way of making the world a tiny bit better.
- Volunteers need coordination, training, and guidance.
- Volunteers need (and deserve) a clear understanding of their role and how it relates to the whole.
- Volunteers should have written agreements and job descriptions.
- Volunteers need and deserve honest feedback.
- Volunteers need praise, recognition, and thanks.
- Volunteer work needs the protection of insurance.
- Volunteers occasionally need to be reassigned or fired.

About Urban Forestry Volunteers

Urban forestry volunteers tend to be older, well respected and influential in the community, knowledgeable about trees and the natural environment, dependable, and social. They are, in general, easy to work with.

Any urban forest manager who plans to complete a project with the help of volunteers should observe the following guidelines.

- Volunteers will probably have to be recruited. Only occasionally will they seek you out. Likely sources of volunteers are neighborhood groups, civic and service clubs, environmental organizations, businesses, industries, and youth groups.
- When considering who to solicit, think first of those who live near the project site and those who might directly benefit. Consider, also, those who might be prone to oppose, damage, or vandalize the project. Ask them to help. Involvement helps build a sense of ownership.
- When asking for volunteers, explain what the project is, why it is important, how long it will take (4 hours should be the maximum to ask volunteers to do physical work), and how it fits into a larger plan for the neighborhood and city.

- When contacting an organization (club, association, industry, other) ask a member to serve as coordinator to distribute information internally and sign up individual volunteers. This relieves you of details and gives you one person to work with.
- When volunteers have agreed to help, give them complete information (in person or through the coordinator) including where, when (including a rain date), and what to wear and bring.
- Take care of all other details—materials, equipment, legal requirements (including liability), on-site communications, traffic signing, first aid materials, sanitary facilities, water, refreshments, and on-site training and supervision. Leave nothing to chance.
- Attend to public relations. If neighbors are influenced, prepare doorhangers or flyers explaining what is going on and asking for their cooperation. Invite the media. Neighborhood improvement projects (particularly tree plantings) are cheerful news.
- On the day of the project, get there early to check on-site details. Welcome each volunteer, assign the volunteers to trainers, and put them to work. Volunteers do not like to stand around wondering what to do. (It is also very bad public relations.)
- As the project progresses, check on everyone and everything.
- When the media arrives, briefly explain what the project is and what it means to the community. Invite reporters to talk with the volunteers. They are the story, not you.
- Thank everyone in person and later by letter. If you can afford them (perhaps they can be donated) give each volunteer a gift commemorating the project. Caps or tee shirts with lettering appropriate to the event are appreciated items. Certificates may also be awarded.
- In writing thank the organizations that volunteers represent—clubs, businesses, industries, other. Thank especially the coordinators.

SECURING FINANCIAL AND TECHNICAL HELP

City forestry departments are not islands unto themselves, depending entirely on finances from the city's general fund and inaccessible to outside sources of technical help. Quite the contrary is true; there is a large network of both financial and technical support available to the city forester.

Financial Help

As would be expected, the majority of city forestry program operational funds are from public monies allocated through the planning and budgeting process. Such public monies are most often from the city's general fund, but they may also come from tax revenues specific to urban forestry, such as frontage tax levies and benefit district assessments. There may also be other tax sources, such as building permit and subdivision assessments, but revenues derived must, in most cases, be applied to specified areas and are most often for tree planting. In cities where monies come form the general fund, urban forestry is competitive with other programs and services of city government. In cities having forestry tax levies, urban forestry is often the function of a separate commission with financial autonomy.

Opportunities for city forestry departments to secure additional funding (beyond city budget allocations) for general operations are rare. In some states, cities may access state highway revenues for maintenance of certain streets and highways, including care of trees and other landscape plants. Additional funding, when available, is for specified purposes, most likely tree planting. In exceptional cases, tree trusts may be established as part of philanthropic tree-planting grants, with earnings applied to maintenance. In a sense, trust earnings may be considered as funds for general operations, but properly are not, as they are intended for maintenance of trees planted under such grants.

Funds provided through recurring budget cycles provide for continuity consistent with planned long-range objectives. Although often useful and even necessary, funds from other sources, earmarked for tree planting, can result in more trees than can be maintained by regular program funds. This problem will be exacerbated as management costs increase over time, unless regular budgets can be increased. Such negatives should not discourage urban forest managers from seeking other funds, however, if there are special needs that will not result in additional unfundable costs over time or if management of streetside trees is the responsibility of adjacent property owners. Excellent examples of special needs are inventories, systems for processing management information, new technologies, and special training for staff or tree board members. There may also be needs for special funding to deal with insect, disease, or storm emergencies.

There are several potential sources of funding for such purposes, as well as for tree planting. These sources—state, federal, and private— are summarized in Appendix 4. Grants provide most funds. To be eligible for grants, cities must show a need and be able to clearly identify benefits that will result from funded projects. Grants will sometimes require in-kind or monetary matching. It is essential that urban forest managers seeking grants understand the grantor's requirements (restrictions on use of grant monies, application procedures, and timetable) and submit proposals clearly showing needs and benefits. Telephone and personal contacts with grantor's representatives can be helpful in obtaining information and conveying a sense of legitimate need. Appendix 5 discusses grant proposals from a grantor's perspective.

A recommended publication for those seeking grants is *Program Planning & Proposal Writing,* Expanded Version, available from The Grantsmanship Center, Dept. DD, P.O. Box 6210, Los Angeles, CA 90014. This publication expands on the following eight elements of grant proposals, giving helpful information on how each should be developed.

1. *Summary.* Clearly and concisely summarizes the request.
2. *Introduction.* Describes the requesting agency's qualifications or credibility.
3. *Problem Statement or Needs Assessment.* Documents the needs to be met or problems to be solved by the proposed funding.
4. *Objectives.* Establishes the benefits of the funding in measurable terms.
5. *Methods.* Describes the activities to be employed to achieve the desired results.
6. *Evaluation.* Presents a plan for determining the degree to which objectives are met and methods are followed.
7. *Future or Other Necessary Funding.* Describes a plan for continuation beyond the grant period and/or the availability of other resources to implement the grant.
8. *Budget.* Clearly delineates costs to be met by the funding source and those to be provided by the applicant or other parties.

An additional source of detailed information about grant proposals and grant availability (particularly from private foundations) is the Foundation Center Library in each state, normally located in the capital city. Telephone numbers of Foundation Centers may be obtained from telephone directory assistance.

Here is a final note about financial help. An often-expressed concern of urban forest managers is that acquisition of grant monies or other funds will have negative impacts on normal budgets—that city administrators, knowing about outside funds, will reduce normal budgets accordingly. Perhaps more fear than reality, this concern may be allayed by meeting in advance with administrators, explaining how the proposal meets a special need without calling for more city funds.

Technical Help

Technical help for urban forest managers, in the form of information and direct assistance, is available from a variety of sources, as summarized in Appendix 5. A primary source is the state forestry agency. Supported by the USDA Forest Service, state forestry agencies have developed increasingly effective urban forestry programs. Services include coordination of urban forestry activities within the state, sponsorship of educational events, guidance in developing new programs, administration of Tree City USA, and direct assistance with application of new technologies. As described previously in this chapter, The Cooperative Extension Service is also a valuable resource. Backed by university specialists in many areas pertaining to urban forestry, local ᴖxtension agents coordinate educational activities, provide publications, and can access research findings on special topics. Other sources are materials suppliers, national arboricultural firms, nurseries, arborists' associations, and national organizations such as The International Society of Arboriculture, The National Arborists Association, The National Arbor Day Foundation, and American Forests.

As alluded to in an earlier section, a rapidly expanding source (or perhaps more properly a method of information access) is through computer information networks. Limited by cost and perhaps user resistance and imagination of application, these systems, while extremely useful at present, hold awesome promise for the future. Urban forest managers not only must keep abreast of current knowledge but also must be closely familiar with new technologies that make it possible.

SUMMARY

Three internal objectives are vital to comprehensive management: (1) a strong city forestry department, (2) effective utilization of external resources, and (3) high-quality performance from those who service the urban forest. Program analysis is an ongoing internal management need and should be based on the question, "Is the program as effective as it should be in meeting the needs of the urban forest?" Citizen support is necessary for effective urban forestry programs and must be cultivated. Urban forestry public relations must be ongoing, through day-to-day activities sensitive to the public, and should also be planned. Indirect management can be accomplished by improving the knowledge and skills of owners and managers, improving the knowledge and skills of those who provide services, and improving the quality and variety of plant materials available to owners and managers. Essential to indirect management are recognition and utilization of information and education opportunities, positive relations with providers of services, and effective use of volunteers. Opportunities for funding in addition to city budgets are available but should be in context with overall program needs and abilities. Technical help is readily available.

REFERENCES

Kincaid, S., Urban Forester. November 1993. From remarks at urban forestry workshop, St. Louis, MO.

Nighswonger, J. J. Fall 1994. *Urban and Community Trees Newsletter.* State and Extension Forestry, Kansas State University, Manhattan, KS.

Society of American Foresters. 1995. "Tips for Dealing With the Media." Communications Department, Society of American Foresters, Bethesda, MD.

National Arbor Day Foundation. 1993. Tree Line *U.S.A.,* Tree City U.S.A. Bulletin No. 25. National Arbor Day Foundation, Lincoln, NE.

CHAPTER *8*

Summary

THE URBAN FOREST

The urban forest involves entire cities and their environs and is an environment of trees and related organisms, structures, and people. The urban forest has many parts, from undisturbed natural woodlands to open areas nearly void of trees. There is a great complexity of ownerships, with responsibility for care of trees sometimes being legally transferred to others. Rights of ownership are not absolute, in deference to the needs of society. The urban forest is also in a constant state of flux, as people continually plant trees, push them aside, build within natural woodlands, and fashion the forest to their needs. The practice of urban forestry is an attempt to manage these activities in such a way as to make trees and related organisms, structures, and people as compatible as possible. (See Figure 8.1.)

THE CONCEPT OF COMPREHENSIVE MANAGEMENT

Comprehensive management involves recognition of the totality and complexity of the urban forest and its relationship to the total urban environment, understanding the needs of the forest, identifying opportunities for positively influencing its care, and applying judicious action. The manifestation of comprehensive management is the direct applica-

FIGURE 8.1. The practice of urban forestry has to do with making trees and related organisms, structures, and people as compatible as possible.

tion of care by a central organization on areas under its jurisdiction and indirect application on all other areas. From the central perspective of the urban forest manager, the total urban forestry environment may be orchestrated with resources applied in context with the needs of the whole. Thus, comprehensive management may be summarized by five basic concepts: the total urban forestry environment must be considered; six elements are necessary for effective management; the needs of the total urban forest must be recognized; appropriate management must be applied, directly and indirectly; and all management is interrelated and must be in context with long-range objectives.

THE REQUIREMENTS

There are six requirements for successful comprehensive management. Each must be in place if the needs of the total urban forest are to be effectively met: a central organization with responsibility and authority; knowledge of the urban forestry environment—biological, institutional/social, and legal; knowledge of the needs of the urban forest; plans for meeting the needs; adequate funding; and effective implementation. In a real sense, effective implementation is the culmination of the preceding five. It deserves equal listing, however, as it is also the measure of success and the key to program continuation.

MANAGEMENT PRACTICES

Accurate information is essential for program planning and implementation. Surveys and inventories provide baseline information for identifying program objectives, determining priorities, and, if continued, guiding annual activities. Types of surveys and inventories and the information gathered must be consistent with program needs.

Urban forestry program planning must be long-range, with objectives and strategies clearly identified. Long-range planning is the function of the city forestry department and tree board, with citizen input. Operational plans are incremental to long-range objectives and are the basis for annual budgets. Plans of work are internal documents containing tasks and timetables. Evaluation is necessary to measure progress and allow necessary adjustments to be made.

Management is manifested by four fundamental practices to meet the needs of the urban forest: tree establishment, maintenance, protection, and removal, with each having separate elements. Tree establishment requires location and selection, planting, aftercare and analysis. The practice of location and selection is critically important, providing the best opportunity for making the future urban forest functional and economical. The highest priority of maintenance must be hazard management, requiring a system for systematic inspection, prevention, and correction. Pruning, as a means of quality and functional improvement, includes hazard reduction, structural development, health management, and aesthetic considerations and is generally the largest single element in urban forestry programs. Protection of the urban forest is accomplished through plant health management and fire protection. Removal needs result from not only dead trees but also other materials that trees naturally discard or that must be disposed of in management operations. Some of what must be removed can be used, depending largely on economic feasibility. Other activities related to urban forest management are providing for trees during construction, managing wildlife, and establishing values of trees.

LEVERAGING EFFORTS

Program evaluation is an essential management activity to determine program needs. Effective urban forestry programs are directly related to public support. Public support is largely a product of public relations—positive day-to-day operations and planned programs to keep trees in the forefront. Working with the media is central to building public support. Program effectiveness (and the quality of the urban forest) can be improved greatly by increasing the knowledge and skills of those who provide services: arborists, nursery people, and others. Volunteers are a major resource, providing labor and other services and serving as advocates of the urban forest. Although normal city budgets constitute the backbone of most urban forestry programs, other funding sources are often used. Technical support for management activities is available from a variety of public and private sources.

Westminster Tree Commission
Annual Report for 1994

MEMORANDUM

P.O. Box 010 - City Hall
Westminster, MD 21158

848-9002
(fax) 876-0299

TO: Tree City U.S.A. Program
FROM: Westminster Tree Commission
DATE: December 23, 1994
RE: Westminster Tree Commission–Annual Report for 1994

Below please find a list of the annual work plan items that were undertaken and completed by the Westminster Tree Commission in 1994.

1. Organized and implemented "Arbor Day Week" activities during the week of April 5th through April 9, 1994. A ceremony recertifying the City of Westminster with Tree City U.S.A. status was held on April 6, 1994, at the Westminster Elementary School. To commemorate this event, 18 trees, consisting of a combination of

sugar and red maple, were planted on the Elementary School grounds. One of the trees was ceremoniously planted as part of the Arbor Day Week proclamation by Mayor Brown. A number of the students assisted the Mayor in the planting of this tree. On other days of the week, either tree plantings or tree pruning was conducted at different sites within the City. The Tree Commission coordinated the site location and species selection for the tree-planting events.

2. To complete the Arbor Day Week activities, the Westminster Tree Commission sponsored the 1994 Carroll County Community Arbor Day Celebration held at Greenway Gardens on Saturday, April 9, 1994. This celebration included the presentation of Tree City U.S.A. certifications and growth awards to the other eligible municipalities in Carroll County. Also, the Maryland Department of Natural Resources presented the Arbor Day Poster Contest award at this event. As part of the Arbor Day celebration, each municipality within the County that was certified as a Tree City U.S.A. donated a tree, which was planted at Greenway Gardens.

3. The City of Westminster was recognized by the State of Maryland as a PLANT Community, and was awarded "Green" status, at the State's Arbor Day celebration on April 9, 1994.

4. In January a presentation on Logging Aesthetics by Geoffrey T. Jones, Director of Land Management of the Society for the Protection of New Hampshire Forests, was cosponsored by the Tree Commission along with the Carroll County Forestry Board, University of Maryland Cooperative Education Service, and the DNR Forestry Service.

5. Prepared and reviewed budget requests that were included in the City's adopted Capital Improvement Program for the next 6-year period.

6. Provided a representative to the West Main Street/Pennsylvania Avenue Reconstruction Task Force, a committee of local citizens, to review and make recommendations for the reconstruction of portions of Pennsylvania Avenue and West Main Street without destroying all existing trees and other historical and cultural features in the area. The Tree Commission representative assisted with formulating decisions on which trees to retain and which to remove.

7. Continuing to revise the draft Hazard Tree Policy as a step in the development of a comprehensive tree management program. It addresses the development of a hazard tree management program for minimizing the exposure of people and property to hazardous tree conditions. The Hazard Tree Policy requires the

establishment of a program that will effectively identify and remove hazardous conditions associated with trees in a reasonable period of time and within the constraints of the City's resources. It was forwarded to the Mayor and Common Council for consideration and has not yet been adopted. However, at the direction of the Mayor and Common Council, the daft document has been revised.

8. Reviewed the plans prepared by the State Highway Administration for landscaping along Route 140 within the City limits. Offered suggestions to include several alternative low-maintenance plants.

9. On the recommendation of the Tree Commission, the Mayor and Common Council entered into a 3-year contract with the Bartlett Tree Expert Company for tree care, maintenance, and Integrated Pest Management.

10. In 1992 the City adopted an Ordinance that prohibits tree topping, unless a permit is granted. The Tree Commission responded to several requests for a permit to top trees. After inspecting the trees, the Commission determined that topping was not appropriate. The Commission offered advice and recommendations to the property owners on how to prune their trees properly and/or to remove the tree instead and replant with an appropriate species.

11. The Tree Commission works with the Town Planner, who staffs the Tree Commission. The Town Planner serves an administrative function for the Tree Commission, assists in developing the Tree Commission policies and in implementing the Tree Commission policies and goals, serves as a liaison between the Tree Commission and the City government, and assists in coordinating tree plantings, Arbor Day celebrations, and the Tree Commission Fall Tree Workshop.

12. The requirements of the Landscape Manual, which requires all developers to include landscaping plans as part of their development plans, is enforced during the plan review process. The Tree Commission aids the Town Planner in the review of landscape plans for private sector projects including but not limited to Popeye's, Marta Technologies, Freewing Aircraft Corporation, Addition to Cranberry Shopping Center, Tevis Oil, Fenby Farm, Peterson Hall Water Main and The Studio at Western Maryland College, Staples Retail Store, Piper Business Campus, Zepp's Hill, Little People's Place, Becker's Auto Center, Fletcher Garage/Apartment, and the Carroll County Health Center.

13. Assisted the Town Planner in conducting inspections of the land-

scaping at the recently completed subdivision of King's Choice, Section 2.

14. Assisted in the City in the resolution of outstanding landscaping issues at the Parr's Ridge Condominium site, in which the developer planted additional trees on the site.
15. Provided guidance to City staff on substitutions for trees species at the townhouse community under development in Avondale Run, Phase II, and the Fairways at Wakefield subdivision.
16. Conducted several hazardous tree inspections. Hazardous trees were removed from the following locations: two along Willis Street, one from the Charles Street Tot lot, one at 24 North Court Street, one at 89 John Street, and one at the Water Tower on Sawgrass Court. Additionally, several City trees were pruned to remove hazardous limbs.
17. Thanks to the efforts and enforcement of the DNR Area Forester, the City received 18 trees as replacements for trees that were removed in violation of the Maryland Roadside Tree Care Law. To correct the violation, the utility company provided 18 Patmore Green Ash trees to the City. The Tree Commission visited several sites with the City and selected the area of South Center Street. The Tree Commission marked each location and coordinated the plantings with the landscape contractor.
18. The Tree Commission was featured in an article in the Maryland Municipal League's newsletter.
19. Established a "Matching Grant Program" for tree-planting projects undertaken by Homeowner's Associations (HOA) for the purchase and planting of trees on common open space property owned by the HOA. The Tree Commission offers technical assistance in addition to the matching funds. This year the Avondale Run HOA participated in this program.
20. Assisted the City in identifying trees damaged by snow and ice storms last winter and an extremely violent storm with high winds this past autumn.
21. Provided technical assistance to City staff in resolving an incident with the State Highway Administration in regard to improper and excessive pruning of street trees along East Main, which occurred during the street reconstruction project. Those trees most severely affected will be replaced by the City, and the State Highway Administration will be invoiced for the associated cost. Additionally, the tree care company went back and conducted corrective pruning under the supervision of the DNR Area Forester.
22. During 1994 several trees were donated to the City. The Tree

Commission assisted in the selection of the species when requested to do so and selected the most suitable areas in which to plant these trees. Included, but not limited to the 10 October glory red maple trees planted along Johahn Drive, 1 pin oak was planted at City Hall, and 2 saucer magnolias were planted at Belle Grove Square.

23. Coordinated with the City's tree expert contractor to provide pruning and elevation to the street trees planted along a two-block section of Main Street, from Longwell Avenue to John Street. This pruning will ensure that these trees will continue to grow in the proper form for street trees.

24. Together with the City, the Bartlett Tree Expert Company, and the Maryland Community Forest Council, the Tree Commission organized and developed their Second Annual Community Forestry Workshop on "Tree Maintenance." The workshop was held on October 19, 1994, at Western Maryland College. Letters of invitation were sent to all counties and municipalities in the State. We had 95 people in attendance representing 34 organizations. Topics included the purpose of pruning and an introduction and overview of the A-300 Pruning Standards, Hazard Tree Identification, a demonstration on the hazards of electric lines, the outdoor field sessions on pruning, arboriculture tools, and tree identification. A manual was also developed specifically for this workshop.

25. Coordinated with the City's tree expert contractor to provide pruning and elevation to the street trees planted along John Street, from Winter's Alley to the edge of the City limits.

26. Thanks to the efforts of the Tree Commission, the City was awarded matching grant funds in the amount of $4,200 under the U.S. Small Business Tree Planting Program for the planting of trees at three separate sites within the city. The executed agreement for the grant was received on November 1, 1994. Therefore, the actual implementation will occur in 1995.

27. Provided coordination and technical assistance to City staff in the drafting of bid specifications and planting plans for planting trees along the portion of East Main Street, which was reconstructed by the State Highway Administation during 1994. This included four field visits to walk East Main Street in order to determine the correct species of tree to plant in each of 106 tree pits installed along the sidewalk, and the appropriate shrubs to plant in the 10 shrub beds.

28. Several of the Liberty Elm trees planted at the City nursery as seedlings by the Tree Commission have reached a suitable caliper

and height for transplanting as street trees. The Tree Commmission identified six of the elm trees to transplant along East Main Street, in the historic area of downtown Westminster.

29. Conducted various inspections of City trees, grounds, and gardens. Identified problems and offered recommendations to the City.

30. Provided Mayor Yowan with recommendations on filling vacancies on the Westminster Tree Commission.

31. Assisted the City Parks Board in establishing a wooded buffer between Jaycee Park and the Whispering Meadows Neighborhood. At the recommendation of the Tree Commission, the City had 60 White Pine trees transplanted from the City's nursery at Eden Farm Pond to the park. The Tree Commission selected the 60 pine trees at the nursery, staked out the planting location for each tree at the park, and assisted in the coordination of transplanting these trees with a local landscaping contractor.

32. Assisted City staff in the implementation of Christmas Tree Recycling Program in Westminster. Citizens who dropped off trees for recycling received a coupon for a free seedling.

33. Assisted City staff in the preparation of applications and documentation that lead to the City's recertification as a Tree City U.S.A., the receipt of a Tree City U.S.A. Growth Award, and the recognition as a Maryland PLANT Community.

Summary of Long-Range Urban Forestry Plan

OBJECTIVES

As representatives of the citizens of the City of _____, the members of the City Tree Board commit to achieving by the year 2010 an urban forest that is safe, healthy, diverse, pleasing to the senses, and functional—an urban forest attractive to birds and other wildlife and a place in which our children may delight and learn.

MEANS

To achieve the preceding objectives, the following means will be employed. The details of each will be included in annual operational plans of the City Forestry Department and will reflect priorities as annually reviewed by the Tree Board and the City Forestry Department.

Safety

As prudent and reasonable, the urban forest will be maintained for the safety of residents and visitors by the following means:

- Identification and removal, where hazards exist, of materials including dead trees, dead or weakened portions of trees, roots, ob-

structions caused by roots such as heaved sidewalks, fruits, and other materials produced by trees.
- Removal of trees or portions of trees interfering with pedestrian or vehicular passage or with line of sight to traffic regulatory signs or signals.
- Removal of trees or portions of trees in contact with high-voltage lines or structures.
- Removal of targets in cases of unique trees.
- Structural pruning and other arboricultural practices.
- Planting of potentially safe and vigorous trees.

Health

Health of the urban forest will be ensured by the following means:

- Implementation of a program of plant health management including: monitoring insect and disease situations, providing selective treatment, and maintaining tree vigor.
- Selection and planting of diverse tree species and varieties potentially free from excessive insect, disease, and environmental damage.
- Timely pruning.

Diversity

Diversity of species, ages, and sizes within the urban forest will be accomplished by the following means:

- Removal of hazardous trees.
- Planting of different species of varying size classes.

Pleasing

Landscape design considerations will be incorporated into the urban forest, particularly with regard to new plantings and other developments.

Wildlife

Provisions for birds and other wildlife will be made in the urban forest by the following means:

- Planting of trees and other vegetation favorable to birds and other wildlife.

- Protection of wildlife habitat during development and construction.
- Retention of dead trees and portions of trees for cavity nesting birds where such retention does not present a hazard.
- Erection of artificial nesting structures.

SCOPE

This plan is for the total urban forest of the City, recognizing that the forest is both publicly and privately owned. The preceding provisions will be put directly in effect on lands for which the City Forestry Department has responsibility. Various city codes will influence implementation on other lands. In addition, a vigorous program to educate property owners and assist them in urban forestry matters will be initiated.

Approved this _____ day of _____, 19____

_____, Chair, City Tree Board

A Basic Urban Forestry
Reference Library

ESSENTIAL

Dirr, M.A. 1975. *Manual of Woody Landscape Plants.* Stipes Publishing Co., Champaign, IL.

Fazio, J. R., and D. L. Gilbert, 1986. *Public Relations and Communications for Natural Resource Managers,* Second Edition. Kendell/Hunt Publishing Co., Dubuque, IA.

Grey, G. W., and F. J. Deneke. 1986. *Urban Forestry,* Second Edition. John Wiley & Sons, New York.

Harris, R. W. 1992. *Arboriculture—Integrated Management of Landscape Trees, Shrubs, and Vines.* Second Edition. Prentice Hall, Inc., Englewood Cliffs, NJ.

Johnson, W. T., and H. H. Lyon. 1991. *Insects That Feed on Trees and Shrubs.* Comstock Publishing Associates, Ithaca, NY.

Miller, R. W. 1988. *Urban Forestry—Planning and Managing Urban Green Space.* Prentice Hall, Englewood Cliffs, NJ.

Moll, G., and S. Ebenreck. 1989. *Shading Our Cities—A Resource Guide for Urban and Community Forestry.* Island Press, Covelo, CA.

Pirone, P. P. 1988. *Tree Maintenance,* Sixth Edition. Oxford University Press, New York.

Shigo, A. L. 1991. *Modern Arboriculture—A Systems Approach to the Care of Trees.* Shigo and Trees Associates, Durham, NH.

Sinclair, W. A., H. H. Lyon, and T. Johnson. 1987. *Diseases of Trees and Shrubs.* Comstock Publishing Associates, Ithaca, NY.

Wendell, W. 1989. *Handbook of Landscape Tree Cultivars.* Prairie Publishing Co., Gladstone, IL.

OPTIONAL

Gerhold, H. D. 1989. *Street Tree Fact Sheets.* School of Forest Resources, Pennsylvania State University, University Park, PA.

Rutledge, A. 1971. *Anatomy of a Park.* McGraw Hill, New York.

Financial and Technical Help for Urban Forestry Programs

Sources of financial and technical help are listed here. It is recommended, however, that a basic urban forestry reference library first be assembled. Appendix 3 lists recommended books, as compiled by a panel of urban forestry practitioners.

SOURCES OF FINANCIAL AND TECHNICAL
ASSISTANCE

* State forestry agencies—Education and technical assistance in cooperation with USDA Forest Service; coordination of state urban forestry activities; administration of funding programs for tree planting and other activities.
* Cooperative Extension Service—Cooperative with USDA and Land Grant Universities, with local offices in most counties. Sponsorship and coordination of educational events; availability of university specialists; publications and other materials; soil testing; consultation.
* State highway departments—Possible funding for landscape maintenance of transportation corridors within cities, depending on state policies.

- State emergency funds—Some states provide funds to help local areas recover from natural disasters, primarily for cleanup, but may include reestablishing trees.
- Federal government—Several federal government agencies including the HUD Community Development Agency and the DOC Economic Development Agency; the Small Business Administration; and the USDA Forest Service (through state forestry agencies as indicated previously) administer grant programs for urban forestry-related purposes. Such programs are published annually by the Office of Management and Budget in the *Catalog of Federal Domestic Assistance.*
- International Society of Arboriculture, P.O. Box GG, Savoy, IL 61874. Tel. (217) 355-9411. Technical conferences; monthly *Journal of Arboriculture;* tree valuation formula; arborist certification program.
- National Arborist Association, P.O. Box 294, Amherst, NH 03031. Tel. (603) 673-3311. Pruning standards; training videos and materials; publications.
- American Forests, P.O. Box 2000, Washington, DC 20013. Tel. (202) 667-3300. Sponsorship of Global Releaf; coordination of urban forest councils; Presentation of National Urban Forestry Conferences; publication of *American Forests* magazine and the *Urban Forestry Forum.*
- American Association of Nurserymen, 1250 I Street NW, Suite 500, Washington, DC 20005. Tel. (202) 789-2900. Publishes *American Standards for Nursery Stock.*
- The National Arbor Day Foundation, 100 Arbor Avenue, Nebraska City, NE 68410. Tel. (402) 474-5655. Sponsorship of Tree City U.S.A. program; annual awards; technical bulletins; workshops and seminars.
- State arborist associations
- State urban forest councils
- National arboricultural firms
- Materials suppliers
- Local people with special knowledge

How to Write Grants—A State Grantor's Perspective*

There are five steps you can take to prepare a better grant proposal:

1. Getting qualified
2. Selling your project
3. Separating yourself from the rest of the pack
4. Selling yourself
5. Packaging

GETTING QUALIFIED

It is not uncommon to disqualify 10% of the proposals admitted. People fail to follow instructions. To write a better grant proposal, pay attention to the following suggestions:

- Read Grant Guidelines thoroughly.
- Follow instructions, FOLLOW INSTRUCTIONS, FOLLOW INSTRUC-TIONS!
- Pay particular attention to minimum requirements such as permits, resolutions, letters of certification, and photos.

*Presented by James R. Geiger, Urban Forester, California Department of Forestry and Fire Protection, August 12, 1992, International Society of Arboriculture, 68th Annual Conference, Oakland, California.

- Look for key words such as *must, shall, will not,* and *are not.*
- If you are not sure, contact the Grant Administrator.

SELLING YOUR PROJECT

Build your proposal around the rating criteria because that's how they are evaluated. If there is a weighting scheme, like a multiplier, be sure to stress the weighted criteria in your proposal.

In order to do the best possible job of selling your project, you must

Do	*Don't*
Understand that only you *really* know your project	Assume that we will know what you mean
Guide us through your project	Wander
Follow the format set forth in the Grant Guidelines	Add a lot of fluff
Make your proposal interesting	Use jargon
Be creative in developing your project as well as your proposal	

SEPARATING YOURSELF FROM THE REST
OF THE PACK

The best proposals involve a basic tree project (tree planting, tree care, or tree protection) along with a combination of community involvement, partnerships with other agencies or corporations, and a significant level of community awareness building.

You need to answer the following questions:

Why is your need unique?
Why can your community benefit from this project more than the next one?

Another technique would be to find a new angle that is creative and unique. Some approaches that have been tried include establishing corporate partnerships, kids programs, and job skill training; finding new ways of doing old things; helping in the aftermath of disasters such as the L.A. riots and the Oakland Hills fire; improving barren school grounds; correcting a deficiency that needs to be addressed.

Look for hints within the Grant Guidelines to describe your project. Often the appendix section of the Guidelines is the place to look for these hints. This is where the pruning standards, nursery standards, and planting standards might be located. You should refer to your knowledge of the use of these standards to indicate to the reader that they are important to you and your organization.

Other tips that will help your proposal stand out from the other proposals include:

- Write in simple, easy to understand language.
- Be brief.
- Be positive.
- Don't make assumptions.

SELLING YOURSELF

When awarding grants, it's crucial that they be awarded to an agency or organization that is capable of accomplishing the proposed project. We look at that part of the proposal as being as important as the project that is being proposed. You could have a great project and really sell us on it but fail to sell us on you and lose out to more highly rated proposals.

Here is how you can improve your rating:

- Tell us about your agency or organization.
- Explain what makes you capable of being able to do this project.
- Outline why are you unique.
- Tell us why is your situation unique—climate, water restrictions, and how are you dealing with it.
- Highlight your ability to provide proper care and maintenance of your project (indicate partnerships).
- Outline your community's tree management program from the community tree board to computerized inventory, or what steps you are taking to get there.
- Discuss your species selection and indicate why they are appropriate and will meet specified standards. Refer to any certification.

The more of these elements you include in your proposal, the higher it will rate. We understand that not all cities are alike and that all tree care programs have different levels of sophistication. The important thing is that you tell us how you do it, how local government and citizen volunteers work together to provide stewardship, and how other partners like corporations are involved.

PACKAGING

The final step in developing your proposal is putting it in an appealing package. The following tips will not only help you submit a better proposal but also help us better understand your proposal.

- Don't depend on the cover letter to explain anything about your project or your organization. The cover letters generally aren't read at all or are not read very thoroughly. Put everything in the body of your proposal.
- Make sure that there are no typographical errors in your final copy. It reflects on your ability to deliver.
- Don't use an expensive and extravagant cover. It often detracts from your proposal rather than enhancing it.
- Employing an unusual format to attract attention puts you at risk of focusing the reader's attention more on form than on content. Stick to a traditional format and if possible follow the guideline suggested in the Grant Guidelines.
- Use a professional-looking cover along with a pleasing, easy-to-read type.
- For ease of reading, add a lot of white space. Break up text by using headings and bolding. Add tabs to separate major sections of your proposal and to tie it back to the table of contents.
- When preparing your budget, be sure to include one column for each of the following: requested amount, agency/organization match, and total project.

Good luck!

Selected Bibliography for Urban
Forest Inventory*

Barker, P. A. 1983. Microcomputer databases for data management in urban forestry. *Journal of Arboriculture* 9(11):298–300.

Bassett, J. R. and W. C. Lawrence. 1975. Status of street tree inventories in the United States. *Journal of Arboriculture* 1(3):48–52.

Burns, G. A. 1986. Urban tree appraisal: the formula approach. *Journal of Forestry* 84(1):18,49.

Chan, F. J. and G. Cartwright. 1979. Tree management aided by computer. *Journal of Arboriculture* 5(1):16–20.

Cramer, D. E., F. J. Deneke, and G. W. Grey. 1976. Computer use in city tree inventories. *Journal of Arboriculture* 2:193–196.

Crossen, T. I. 1989. The management of urban street trees using computerized inventory systems. *Journal of Arboriculture* 15(1):1–6.

Evans, W. J. 1981. Reduction of public tree liabilities. *Arboricultural Journal* 7(8):219.

Gerhold, H. D. and C. J. Sacksteder. 1979. TRESYSTM: Tree records system for municipalities. *Journal of Arboriculture* 5(11):256–261.

Gerhold, H. D., K. C. Steiner, and C. J. Sacksteder. 1987. Management information systems for urban trees. *Journal of Arboriculture* 13(10):243–249.

Graham, S. 1987. Counting on trees. *Journal of Arboriculture* 12:91–93.

Grainger, R. D. T. and P. Thompson. 1981. Computer-assisted tree management. *Arboricultural Journal* 7(4):301–308.

*Prepared by Dr. Kim D. Coder, The University of Georgia.

Green, Thomas L. 1984. Maintaining and preserving wooded parks. *Journal of Arboriculture* 10(7):193–197.

Hickman, G. W., J. Caprile, and E. Perry. 1989. Oak tree hazard evaluation. *Journal of Arboriculture* 15(8):177–184.

Radd, A. 1976. Trees in towns and their evaluation. *Arboricultural Journal* 3(1):16.

Richards, N. A. 1979. Modeling survival and consequent replacement needs in a street tree population. *Journal of Arboriculture* 5(11):251–255.

Rodgers, L. C. and M. K. Harris. 1983. Remote sensing survey of pecan trees in five Texas cities. *Journal of Arboriculture* 9(8):208–213.

Sacksteder, C. J. and H. D. Gerhold. 1979. A guide to urban tree inventory systems. School of Forest Resources Research Paper #43, Penn. State University.

Sanders, R. A. and J. C. Stevens. 1984. Urban forest of Dayton, Ohio: A preliminary assessment. *Urban Ecology* 8:91–98.

Smiley, E. T. and F. A. Baker. 1988. Options in street tree inventories. *Journal of Arboriculture* 14(2):36–42.

Talarchek, G. M. 1987. The indicators of urban forest condition in New Orleans. *Journal of Arboriculture* 13(9):217–224.

Tate, R. L. 1984. Municipal tree management in New Jersey. *Journal of Arboriculture* 10(8):229–233.

Tate, R. L. 1985. Uses of street tree inventory data. *Journal of Arboriculture* 11:210–213.

Thurman, P. W. 1983. The management of urban street trees using computerized inventory systems. *Arboricultural Journal* 7(2):101–117.

Valentine, F. A., R. D. Westfall, and P. D. Manion. 1978. Street tree assessment by a survey sampling procedure. *Journal of Arboriculture* 4(3):49–57.

Wagar, J. A., and E. T. Smiley. 1990. Computer assisted management of urban trees. *Journal of Arboiculture* 16(8):209–215.

Webster, B. L. 1978. Guide to judging the condition of a shade tree. *Journal of Arboriculture* 4(11):247–249.

Weinstein, G. 1983. The Central Park tree inventory: A management model. *Journal of Arboriculture* 9(10):259–262.

Ziesemer, A. 1978. Determining need for street tree inventories. *Journal of Arboriculture* 11:210–213.

Index

Advocacy, 6, 7, 108
Aftercare, 78
American Association of Nursery-
    men, 72, 75, 76
American Forests, 120, 127
*American Standards for Nursery
    Stock,* 72
Analysis:
  program needs, 105–107
  tree establishment, 78
Arboriculture, 57
Arborists:
  certification of, 114
  selecting, 115–117
  utility of, 117
  working with, 113–115, 117–118

Boulevards, 13, 15
Budgets, 52, 125

Cemeteries, 12, 15
Citizens' organizations, 30
City buildings, 12, 13

City forestry department:
  authority of, 6, 22
  function of, 5, 8, 10
  and relationship with other city de-
      partments, 28
Climate adaptability, 62. *See also*
    Cold hardiness
Cold hardiness, 62–63
Commercial areas, 25
Computer applications:
  information management, 102–103
  inventories, 42
  tree selection, 71
Construction:
  tree protection related to, 93
    alternatives, 95
    approaches, 96
    planning, 96
Contracting:
  determinations, 74, 88
  for planting stock, 75
  for pruning, 88–90
  for tree planting, 76
Cooperative Extension Service, 29,
    30, 113, 127

Council of Tree and Landscape Appraisers, 101
Cover-type mapping, 36

Developers, 29, 96
Development, 97

Easements, 22. *See also* Utilities
Excavations, 74

Financial help, 124–126. *See also* Funding
Fire protection, 97
Foundation centers, 126
Funding:
  grants, 125
  sources for, 124–126, 145–146

Geographic Information Systems (GIS), 36
Grants:
  proposals, 126, 147–150
  sources for, 126
Grantsmanship Center, 126
Groundskeepers, 29

Homeowners associations, 33

Industrial areas, 25–26
Information dissemination:
  methods, 112
  principles, 112
International Society of Arboriculture, 82, 101, 102, 127
Inventories, 40–43

Kansas Arborists Association, 113

Landscape architects, 29
Landscape designers, 29

Liability, 33. *See also* Tree hazards

McAlester Tree Board, 6–7
Maintenance, 79–90
  hazard management, 79–85
  plant health management, 85–86
  quality improvement, 96
  removal and utilization, 90–93
  special considerations, 93–102
Management:
  audiences, 10
  authority, 2, 22
  concept of, 2, 130
  direct, 2, 5, 10, 36
  indirect, 2, 5, 10, 37
  of information, 102
  needs of the urban forest, 57, 131
  orchestration of, 2, 7
Media:
  interviews with, 120–121
  recommendations for working with, 119–120
  working with, 119–121
Monitoring, *see* Aftercare
Monuments, 12, 14

National Arborist Association, 91, 120, 127
National Center for Forest Health Management, 86
National Institute for Urban Wildlife, 100
Negative characteristics of trees, 70
Nurseries:
  sources of planting stock, 72
  working with, 118–119

Ordinances:
  landscape, 32
  management, 30
  tree protection, 30

Parking lots, 18, 21
Parks, 10, 11, 19

Planners, 29
Planning:
  benefits of, 45
  criteria for, 46
  development and construction, 97
  emergency, 53–54
  evaluation, 53
  flow, 46
  long-range, 46–51
  operational, 51–53
  public relations, 110
Plans, *see* Planning
Plans of work, 53
Plant health management, 85–86
Planting stock:
  characteristics of, 72–73
  contracting for, 75
  handling, 73
  sources for, 72, 73
  standards of, 72
Politics, 33, 107–108
Pollution:
  ozone, 68
  salt, 61–62
  sulphur dioxide, 67
Program analysis:
  macro, 106
  micro, 106–107
Pruning:
  contracting, 88, 91
  cycles, 87
  management approaches, 87
  priorities, 87–88
  standards, 89–90
Public lands, 22
Public relations:
  opportunities, 110–111
  planning, 110
  principles, 109
  process, 109
Public squares, 10

Regulations:
  administrative, 33
  state and federal, 33
  subdivision, 32

Removal and utilization, 90–93
Residential areas, 22
  characteristics of, 22, 24, 25
  factors influencing the urban forest of, 23
Riparian areas, 19, 21
Rock Creek Park, 19

Salt tolerance, 61–62. *See also* Pollution
Seed sources, 73. *See also* Planting stock
Shigo and Trees Associates, 82
Sidewalks, *see* Treelawns
Site analysis, 59–60
Site constraints, 60–62. *See also* Tree Selection and Location
Society of American Foresters, 86, 120
Southern Trees, 71
Special areas, 20, 23
Staking, 77
State forestry agencies, 29–30, 127
Streetside(s):
  landscape design of, 18, 20
  responsibility for trees on, 13
  tree situations, 13–18
  use of, for utilities, 18
Subdivision regulations, 32–33
Surveys, 37–40

Teachable moment, 110, 112
Technical assistance, 126–127, 145–146
The National Arbor Day Foundation, 118, 120, 127
Tolerance:
  exposure, 63
  flooding, 66
  insects and disease, 67
  pollution, 67, 68
  shade, 63, 64–65
  soil, 63
Tort law, 33, 79. *See also* Tree hazards

Tree boards:
  activities of, 108–109
  authority of, 6
  responsibilities of, 6–7
  roles of, 6
Tree City USA, 127
Tree condition classes, 38, 39
Tree establishment, 58–78
  priorities of, 59
  rules of, 58
  steps in, 59
Tree hazards:
  liability, 33, 79
  management of, 79–85
Treelawns, 13–14, 16–18. *See also*
  Streetside(s)
Tree Line USA, 118
Tree planting, 72–78
  assurance of quality planting stock,
    72
  contracting, 73
  implementation, 76
  site designation, 74
Tree selection and location:
  constraints of:
    local environmental factors, 61
    negative characteristics of trees,
      70
    soil, 61
    spatial, 60
  economics of, 70
  factors in, 59
  other considerations in, 70
Tree size and form, 68–69. *See also*
  Tree selection and Location
Tree topping, 113–114, 115–116

Undeveloped areas, 26–27
Urban forest:
  dynamics of, 2
  management:
    needs, 6, 57
    requirements, 3

  responsibility for, 9
  management inventories, 40–43
  scope, 1
  surveys, 37–40
  valuation methods, 100–101
Urban forestry:
  definition of, vii–viii
  environment of, 9
    economic, 24
    institutional and commercial, 27–
      30
    legal, 30–33
    physical, 9–27
    political, 33–34
USDA Forest Service, 71, 82, 86,
  120, 127
USDI Fish and Wildlife Service, 100
Utilities, 18, 19, 22, 42, 60, 74

Valuation:
  formula, 101
  methods, 100
Volunteers:
  group characteristics of, 122
  guidelines, 123–124
  individual characteristics of, 123
  working with, 121–124

Westminister Tree Commission, 7,
  133–138
Wildlife:
  advantages of urban environment to,
    99
  disadvantages of urban environment
    to, 99
  management:
    aspects of, 98
    assistance with, 100
    methods of, 100
    opportunities in, 99
    principles of, 100